Freedom of Expression and the Internet

by Terri Dougherty

LUCENT BOOKS
A part of Gale, Cengage Learning

GALE
CENGAGE Learning

Detroit • New York • San Francisco • New Haven, Conn • Waterville, Maine • London

LIBRARY OF CONGRESS CATALOGING-IN-PUBLICATION DATA

Dougherty, Terri.
  Freedom of expression and the internet / by Terri Dougherty.
      p. cm. -- (Hot topics)
  Includes bibliographical references and index.
  ISBN 978-1-4205-0227-5 (hardcover)
  1. Internet--Social aspects. 2. Technological innovations--Social aspects.
  3. Freedom of expression. I. Title.
    HM851.D678 2010
    323.44'302854678--dc22
                                                              2010001547

Lucent Books
27500 Drake Rd.
Farmington Hills, MI 48331

ISBN-13: 978-1-4205-0227-5
ISBN-10: 1-4205-0227-1

Printed in the United States of America
1 2 3 4 5 6 7 14 13 12 11 10

Printed by Bang Printing, Brainerd, MN, 1st Ptg., 06/2010

CONTENTS

FOREWORD

Young people today are bombarded with information. Aside from traditional sources such as newspapers, television, and the radio, they are inundated with a nearly continuous stream of data from electronic media. They send and receive e-mails and instant messages, read and write online "blogs," participate in chat rooms and forums, and surf the Web for hours. This trend is likely to continue. As Patricia Senn Breivik, the former dean of university libraries at Wayne State University in Detroit, has stated, "Information overload will only increase in the future. By 2020, for example, the available body of information is expected to double every 73 days! How will these students find the information they need in this coming tidal wave of information?"

Ironically, this overabundance of information can actually impede efforts to understand complex issues. Whether the topic is abortion, the death penalty, gay rights, or obesity, the deluge of fact and opinion that floods the print and electronic media is overwhelming. The news media report the results of polls and studies that contradict one another. Cable news shows, talk radio programs, and newspaper editorials promote narrow viewpoints and omit facts that challenge their own political biases. The World Wide Web is an electronic minefield where legitimate scholars compete with the postings of ordinary citizens who may or may not be well-informed or capable of reasoned argument. At times, strongly worded testimonials and opinion pieces both in print and electronic media are presented as factual accounts.

Conflicting quotes and statistics can confuse even the most diligent researchers. A good example of this is the question of whether or not the death penalty deters crime. For instance, one study found that murders decreased by nearly one-third when the death penalty was reinstated in New York in 1995. Death penalty supporters cite this finding to support their argument that the existence of the death penalty deters criminals from committing murder. However, another study found that states without the death penalty have murder rates below the national average. This study is cited by opponents of capital punishment, who reject the claim that the death penalty deters murder. Students need context and clear, informed discussion if they are to think critically and make informed decisions.

The Hot Topics series is designed to help young people wade through the glut of fact, opinion, and rhetoric so that they can think critically about controversial issues. Only by reading and thinking critically will they be able to formulate a viewpoint that is not simply the parroted views of others. Each volume of the series focuses on one of today's most pressing social issues and provides a balanced overview of the topic. Carefully crafted narrative, fully documented primary and secondary source quotes, informative sidebars, and study questions all provide excellent starting points for research and discussion. Full-color photographs and charts enhance all volumes in the series. With its many useful features, the Hot Topics series is a valuable resource for young people struggling to understand the pressing issues of the modern era.

A New Era for Free Speech

The freedom to express an opinion is a right Americans have long treasured. Freedom of speech is guaranteed to Americans under the First Amendment. The advent of the Internet has given Americans another way to use this freedom but has also ushered in new challenges to the free-speech guarantee.

The U.S. Congress has enacted laws that place limits on speech online. For example, Federal Trade Commission regulations require bloggers to disclose when they receive payment or goods for writing about a product or reviewing a book, a restriction that print journalists do not have. In addition, online speech is still governed by laws related to defamation of character, libel, or slander. A derisive comment made online takes on a level of gravity it would not have had if it were spoken in private to a group of friends. For example, fashion designer Dawn Simorangkir sued rocker Courtney Love for libel after Love posted uncomplimentary posts on Twitter about her. Among other things, Love called the designer a liar and thief. Love said she had the right to make the statements because of her free-speech rights, but Simorangkir and her lawyers disagreed and disputed Love's argument that her comments were protected. Love asked that the court dismiss the case, but a judge refused and the case remains in court.

Even a seemingly offhand comment can lead to conflict. A woman who sent a Twitter message describing her condo as moldy was sued by her landlord, who said the comment was libelous.

The online speech rights of students are especially blurry. Students who post online comments about their teachers, friends, or school administrators face consequences that would likely not have arisen if they had been simply talking with their friends at home or writing in a personal journal. A critical comment about a teacher or friend that is posted online, even when it is written on a home computer, could result in disciplinary action at school or even lead to a court case. One student who created a fake social networking site that parodied his teacher was sued for defamation when the teacher claimed his comments were libelous. In another case a court ruled that a school had the right to discipline a student who wrote about a disagreement with a school administrator in an online blog. Even though she wrote the comment on her home computer, the court decided she could be disciplined for using language the school did not feel was appropriate.

Rights and Limits

While the Internet has raised questions about how far a person's free-speech rights extend, in many ways a person's right to free speech online is given strong protection. The U.S. Supreme Court has ruled that debate over "public issues should be uninhibited, robust, and wide-open."[1] This right extends to the Internet.

However, the right to free speech, both online and elsewhere, is not without some limits. Threats, obscenity, and libelous statements are not allowed. Child pornography is illegal online, just as it is elsewhere. Copyright laws have had to be updated to cover Internet technology that makes it easy to copy a song or page from a book, but the rules that protect ownership of music, writing, and images are enforced on the Internet as well as in print or on video.

What constitutes an unacceptable online comment and what is protected speech can be a matter of contention, however. It is not always clear which statements are allowable and which are not. Former Alaskan governor and vice presidential candidate Sarah Palin challenged comments made by a blogger and threatened him with a lawsuit after he posted an item claiming she was

Online free speech is strongly protected, but libelous comments are prohibited, even on social networking sites like Facebook.

getting a divorce. In another case celebrity blogger Perez Hilton said actress Demi Moore defamed him after she said on Twitter that he was disobeying child pornography laws after Hilton posted a link to a photo of Moore's fifteen-year-old daughter wearing a low-cut blouse.

Changing World

What types of speech should be restricted or classified as illegal continues to be questioned as new forms of communication add additional layers to the free-speech debate. Restrictions have been placed on what can be said on the radio, for example, with obscene words being outlawed because words broadcast over the airwaves come into a person's environment so quickly that they cannot be stopped. Broadcast television programs also fall under censorship regulations. What level of oversight or restriction, if any, should be placed on Internet communication is a free-speech issue that began to erupt when the World Wide Web came into popular use in the last decade of the twentieth century.

Making a statement online can have repercussions, as bloggers hit with libel suits have found and students have learned when disciplined at school for rebellious online comments. Freedom of speech allows the Internet to be a forum for open and opinionated discussion, but also brings with it a responsibility to act within the law. The limits of the law continue to develop.

CHAPTER 1

FREE SPEECH AND RESPONSIBILITY

The Internet has become the medium of choice for people who want to let their feelings, thoughts, and ideas be known in a public forum. It offers individuals the ability to send information more quickly and to a broader audience than any previous means of communication. The nature of the Internet allows diverse opinion and content to be rapidly accessible to a large group of people.

Internet technology takes away the barriers that used to make publishing or broadcasting an opinion or a piece of news something that only a select group of people had the ability to do. The media traditionally had gatekeepers in the form of reporters, editors, and publishers who decided which issues to write about, what to say, and when to publish it. Television news producers had control over what was broadcast, and radio show hosts determined who spoke over the air. With the Internet, a person no longer needs access to a newspaper column, radio show, or television program to share his or her views widely. Access to a social networking site or an online blog is all that is needed to publish an opinion that anyone can see.

Varied Viewpoints

Because the Internet is available to anyone with a computer and the ability to connect with others online, it is a place where many different viewpoints are presented. "The Internet is rapidly emerging as the dominant means of mass communication—transforming traditional broadcasting and cable with new business models and decentralizing the tools of speech and commerce in the information society to all citizens,"[2] noted Ben Scott, policy director for the media reform organization Free Press.

The Internet allows a person to comment immediately on a news story or post an opinion in an online journal or blog. A group in favor of an issue can build a Web site expressing its views, and a group against the issue can build one as well. The Internet levels the playing field by offering an equal opportunity to be heard.

Free-Speech Limits

While the Internet offers an easily accessible platform for spreading information and offering opinions, it is not without rules. Those who take on the responsibility of communicating online must do it within the limits of the law.

Communicating electronically does not give a person the right to say or write things that would not be allowed offline. A

The Internet offers an easily accessible, level platform in which people can express their viewpoints.

Those who communicate online must take on the responsibility of free speech within the limits of the law.

person's right to freedom of speech is balanced with another's right to be safe. Abusing a person by sending harassing letters is illegal, and sending harassing e-mails is illegal as well. Trying to intimidate someone by sending threatening electronic messages is not allowed under the law, just as it is illegal to send threatening letters or threaten someone in person. It is also against the law to send an obscene e-mail. Child pornography is not protected by the First Amendment and is illegal on the Internet as well.

Defamatory statements or lies that can destroy a person's reputation cannot be posted on the Internet without risk of generating a lawsuit. Commentators, bloggers, and tweeters can be sued if the information they post is defamatory. When a defamatory statement is written, it is called libel. When it is spoken, it is called slander. Both libel and slander are considered harmful, and a person can be brought to court for making that type of statement.

Public and Private Standards

The First Amendment protects a person's right to discuss vigorously the actions of public officials and others in the public eye. When it comes to defamation and the law, different standards apply for public and private people. A public person is someone who is often in the news, such as an actor or politician. A private citizen could be a neighbor, friend, or relative.

DEFAMATION GOES TOO FAR

"People have the right to free speech. But they've never had the right to defame someone. They still don't."—Matt Zimmerman, senior staff attorney at the Electronic Frontier Foundation.

Quoted in Rebecca Webber, Online Comments Spark Lawsuits, *Parade*, September 20, 2009. www.parade.com/news/intelligence-report/archive/090920-online-comments-spark-lawsuits.html.

Criticizing political figures or others in the public eye can usually be done without fear of a libel suit. For example, if a story is printed that contains an error about a politician, the politician must prove that the person who wrote the story intended to make the error and made it in order to harm the politician. This is very difficult to prove.

The right of journalists to make comments about political figures is granted to newspaper reporters and broadcasters, but courts have extended this right to people who post information on the Internet as well. For example, a Minnesota man who published a blog called Minnesota Democrats Exposed was protected by this right after he was sued for libel after posting a critical comment about a Democrat. A judge dismissed the case, saying that the person who had been criticized was a public figure and had to prove that the statement was made recklessly or with malice.

Dangerous Lies

Although defamation can be difficult for a public person to prove, making up a harmful statement about someone and putting that false statement online is clearly outside the protection of the First Amendment. A person who knowingly posts a false statement that defames someone could be sued in civil court, which handles cases that fall outside criminal law, such as damage to reputation or contract disputes.

Because defamation harms a person's reputation, a person who believes someone else has made a libelous or slanderous statement about him or her can sue in civil court. The person can ask that he or she be awarded money because of the damage the statement caused. This is what a Georgia lawyer did after a man wrote on his blog that the lawyer had given bribes to judges. This was an untrue statement, and the lawyer sued the blogger in civil court. Because the blogger had posted information that was false and that harmed the lawyer's reputation, a jury found the blogger guilty of defamation. The lawyer was awarded fifty thousand dollars.

Video Defamation

A defamatory statement does not necessarily need to be a blog entry, MySpace page, or a written online comment to bring up questions of defamation. What is said and posted in a video can also pose a problem. Two brothers learned this when they put their thoughts about a summer job at an A&P grocery store to music and created rap songs such as "Produce Paradise" and "Always Low Prices" while performing under the name Fresh Beets. They filmed a "Produce Paradise" video in the store and posted it on YouTube and their Web site. They were sued for $1 million by the Great Atlantic and Pacific Tea Company, which owns A&P. The case was settled out of court, and the video was taken off the Internet.

In another case a former student at the University of North Dakota was sued after posting untrue information about her professor online. The student used an Internet site to post information about the professor, saying he had exchanged sexually explicit e-mails with her. The professor said this was false and that her Web site damaged his reputation. He sued her for defamation. The court agreed with the professor and gave him $3 million in damages.

DANGEROUS WORDS

"First Amendment or no First Amendment, sometimes speaking freely—and publicly—about other people or organizations comes with the threat of a lawsuit if you criticize them or divulge information they don't want revealed. It can also get you fired."—Writer Reid Goldsborough.

Reid Goldsborough, "Blogging and the Law," *Information Today*, September 2005, p. 32.

Fact and Opinion

While a person posting information on the Internet must be careful not to defame someone, that does not prohibit a person from expressing a strong opinion online. Defamatory statements are false statements of fact, which can be proved to be true or false. An opinion is a person's view of another person or event. When the Georgia man said that the lawyer had taken bribes, he was making a statement of fact. His assertion could be proved true or false by checking with the local judges to see whether the bribes had been made or not. In that case the statement was proved to be false.

A statement of opinion cannot defame someone. A person can comment that a teacher is the best teacher in the world or the worst teacher in the world, and that statement of opinion is protected by the First Amendment. "Things that are not provable true or false are not defamation," explains Mary-Rose Papandrea, a member of the Boston College Law School faculty. "If you say someone is the worst teacher ever, that's not actionable. If you say the teacher hit a student, that's different, that's provable true or false."[3]

Using an Assumed Name

People are also at risk for a defamation lawsuit if they lie about who they are and post untrue information under an assumed name. A person cannot assume the identity of someone else and post false comments without risking a lawsuit. This was made clear to a group of high school students who posted a fake profile of their high school principal on MySpace. The profile said the principal went to strip clubs and was a member of the Ku Klux Klan. The principal sued several students for posting the fake profile. The students who created the profile admitted to

On sites like MySpace, people can risk a defamation lawsuit by posting untrue information, even if it is posted under a false identity.

what they had done and were punished by the school, and the lawsuit was dropped. While the case did not continue in court, it shows the consequences that can result when a person posts lies under another's name.

A Pennsylvania student was also the subject of a defamation lawsuit after he posted a controversial parody about his principal on MySpace. The profile included profanity and alleged that the principal had beer in his desk and used marijuana. The student was disciplined by the school for posting the information and later filed a lawsuit that claimed he should not have been punished because he had a First Amendment right to post the parody. However, the principal also sued and claimed that the site defamed him. The case, which remains unsettled, illustrates the questions that arise when one person uses his free speech right to create a parody but another claims to have been harmed by it. Parody is not necessarily illegal, Papandrea notes, but problems emerge when someone looks at the site and mistakes it for a real profile.

Stating the Truth

It is permissible to post true information, even if that information hurts a person's reputation. If a person has been convicted of a theft, it is not defamatory to talk about his conviction online. If a student has a video proving that a teacher sleeps during class, it is not defamatory to say that the teacher sleeps during class. If a statement can be proved to be true, it is not defamatory.

A person making a statement on the Internet risks a lawsuit, however, if he or she repeats information that has not been verified as true. This issue was brought to court when sportswriter Jonathan Givony was sued by Joel Bell, an agent for National Basketball Association players, after Givony wrote that Bell was "an extremely sketchy agent"[4] and accused him of paying bribes to the parents of players he was hoping to sign. Givony later admitted that his comments had been out of line and that he had written his story after hearing information about Bell's reputation. Bell said that the statements were defamatory and filed a lawsuit requesting $25 million in damages. The Media Law Resource Center reported that the case was settled for an apology and ten thousand dollars.

Responsibility and Freedom

Making an untrue allegation can be costly and underscores the responsibility that comes with the freedom of speech. Americans are guaranteed the right to state what is on their minds but also

Companies create policies for their employees with regard to the appropriate use of Internet communication while at work or while using company computers.

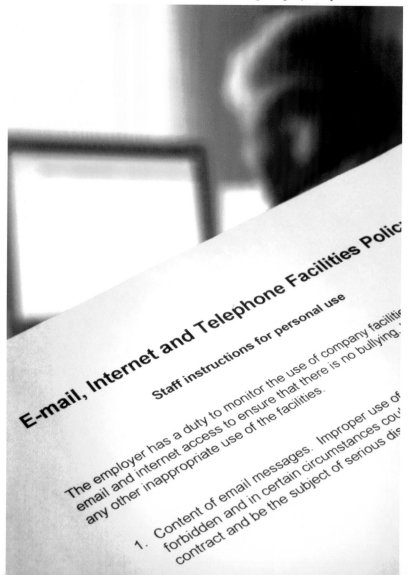

Hoax Victim Sues AOL

Kevin Zeran was the victim of an Internet hoax and sued his Internet service provider for allowing it to persist. Someone posted his name and phone number on an America Online (AOL) bulletin board and said that he was selling T-shirts and other items that mocked the Oklahoma City bombing of the Murrah Federal Building, in which 168 people were killed and 850 were injured. After Zeran began getting death threats and nasty phone calls, he complained to AOL. The company removed the posting but did not print a retraction. The notices kept appearing on the board for several weeks, although Zeran asked AOL to block them. In 1997 Zeran sued AOL for being negligent. The court ruled, however, that AOL was not responsible for what was said on its site. The huge amount of information transferred over the Internet meant that making AOL responsible for what was said on its site had the prospect of chilling free speech.

have a right to protect themselves through the court system if someone makes a defamatory statement about them.

The Internet provides a new format for exercising the right to free speech but carries with it the same limits that exist in the offline world. It has raised some new questions as well. As a relatively new means of communication, the Internet brings a new dimension to free-speech issues. Its far-reaching nature raises questions over the boundaries of the law. Lawmakers, government leaders, and private citizens wrestle with the question of what can and should be said in the online environment.

OPEN ONLINE DIALOGUE

Like the Internet, newspapers, books, radio, and television are all media that people use to share information and are places where a person can make a statement or express an opinion. Although all are used for communication, they do not all receive the same level of free-speech protection. The programs that are broadcast on television and radio receive additional scrutiny and are regulated by the Federal Communications Commission (FCC), but newspapers and books are not regulated by a governing body. Laws govern libel, slander, threatening speech, harassment, and child pornography, but additional regulations apply to information that is broadcast.

Part of the debate over freedom of speech on the Internet has centered on how the medium should be defined. The Internet can be used to present ideas in a written format as newspapers and books do, but it also has video and sound capabilities like radio and television. Clearly, then, the Internet has components of many types of media, yet its immediacy makes it even more difficult to regulate. Something can be posted on the Internet instantaneously and can just as easily be withdrawn.

Regulation Versus Freedom

The Internet is not the first new medium to raise questions about regulation. When movies were introduced in the early 1900s, the Supreme Court ruled that they did not deserve the same protections as other speech because moving pictures were a business that was not to be thought of as part of the press. Movies were a representation of events, the Supreme Court ruled, presented in an attractive manner that could be used for evil. Therefore, the court ruled that communities had a right to censor them. This meant that community review boards could hold back movies they thought were inappropriate, perhaps because of obscenity or sexual references, and refuse to allow theaters to show them. This view changed, however, and the court ruled in the 1950s that movies did indeed communicate ideas and deserve First Amendment protection. Although the criteria used by review boards were impacted by this decision, movies continued to be subject to censorship by review boards for several decades after that. The film industry uses a self-governing rating system to alert viewers to movie content using labels such as G for general audiences, PG for parental guidance suggested, and X for movies that are inappropriate for children.

While movies did eventually receive full First Amendment protection, some limits are placed on television and radio broadcasts. Because television and radio broadcasts come into a person's home so quickly that a parent does not have time to block a child from hearing obscene words or seeing obscene images, the stations broadcasting the material must censor that type of lan-

Early Support

The Supreme Court's decision that struck down the majority of the Communications Decency Act granted broad free-speech rights to Internet users. It also came at a time when the Internet was still a relatively young means of communication. Compared to the decades it took for cable television and movies to receive free-speech rights, the Internet's free-speech status came quickly. "New media are usually born in captivity, and the Court takes a great deal of time—usually decades—before recognizing that the First Amendment applies, much less that full protection is appropriate," noted lawyer Robert Corn-Revere on the First Amendment Center Web site. "Here, rather than presuming that the Internet should receive less protection, the Court held that full First Amendment protection applies unless the government can prove otherwise."

Robert Corn-Revere, "Internet & First Amendment Overview," First Amendment Center. www.firstamend mentcenter.org/speech/internet/overview.aspx.

guage or those images from their programming before it is sent out and children have a chance to see or hear it. Stations that do not abide by these rules face a fine from the FCC. The power of the FCC was affirmed when a radio station aired a monologue by George Carlin called "Seven Words You Can't Say on Radio" in 1973. The Supreme Court ruled that the FCC could forbid it from airing when children would be likely to hear it. The FCC also has the right to fine television broadcast stations if they violate decency standards. For example, CBS television was fined after Janet Jackson's breast was inadvertently exposed to viewers during a 2004 Super Bowl half-time performance.

A NEW ERA

"In countless ways, the Internet is radically enhancing our access to information and empowering us to share ideas with the entire world. Speech thrives online, freed of limitations inherent in other media and created by traditional gatekeepers."—The Electronic Frontier Foundation.

Electronic Frontier Foundation, "Free Speech." www.eff.org/issues/free-speech.

While broadcast television and radio are regulated because they come into people's homes without an opportunity for the viewer or listener to censor what is said, cable television has different standards. Because parents can block certain channels from coming into their home, cable television channels are not subject to the same type of regulation as broadcast channels. This freedom was not given to cable television right away, however, but gradually evolved as a series of cases wound their way through the court system.

The Internet's Turn

Given the history of new media regulation, it was not a surprise to Internet free-speech advocates that questions about Internet regulation arose as the communications medium gained in popularity. "We thought it was pretty clear from the begin-

One of the first challenges the Internet faced was balancing free speech rights with the desire to shield children from inappropriate material.

ning that speech was going to be a big issue on the Internet," said Lee Tien, senior staff attorney for the Electronic Frontier Foundation, which was established in 1990 to protect Internet freedoms. "It's always the case that new means of expression get into trouble. Everyone sees them as the wild, wild west."[5]

Whether the Internet should be allowed the same First Amendment protection as published material or the narrower rights of broadcast material was one of the first issues that arose. Several challenges to Internet free speech have played out in federal and state courts, and most have provided broad protections for Internet users.

One of the first questions lawmakers grappled with was balancing a person's right to freedom of speech with the desire to protect children from inappropriate content. The Internet makes it easier for children to see indecent or sexual material, for example. Lawmakers first imposed regulations on the Internet with the Communications Decency Act, which was part of the Telecommunications Act of 1996. The act prohibited a person from posting indecent or offensive material on a Web page or in an Internet chat room and was designed to keep sexual images such as pornography away from children. Although the law was designed to keep minors from viewing pornography, opponents argued that the words "indecent and offensive" were too broad. This could impact the online discussion of books such as *The Catcher in the Rye* or the seven words that could not be said on radio, noted organizations such as the Center for Democracy and Technology. The American Civil Liberties Union (ACLU) and others who said the act infringed on free speech challenged the act in court. "We began worrying immediately whether the Internet would receive the broader protection of books and magazines or the lesser protections of broadcast TV,"[6] said Chris Hansen, a senior staff attorney with the ACLU.

One of the challenges in the case was the fact that many of the judges involved had never been on the Internet, Hansen noted, and were not familiar with what it was. Because the Internet involved using a screen, like broadcast television, the ACLU was concerned that the judges would compare it to radio and television, where images, words, and sounds could not be easily censored by the user. "We spent a lot of time explaining alternative methods to protect children,"[7] Hansen said, noting that parents could use filtering software to keep unwanted material away from children.

In 1997 the case of *Reno v. American Civil Liberties Union* went to the Supreme Court, where the judges unanimously ruled that the majority of the act was unconstitutional. The court ultimate-

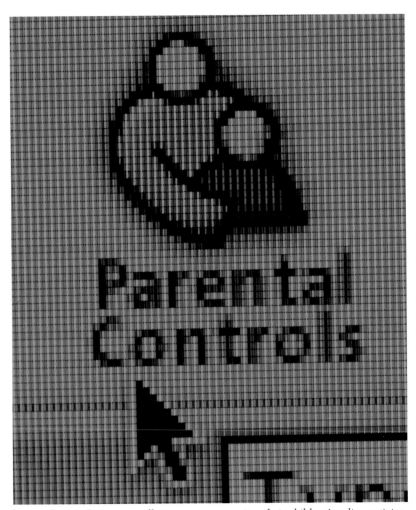

Parental control programs allow parents to monitor their children's online activity or block certain types of content altogether.

ly ruled that the Internet gives parents the opportunity to install filtering devices to protect children, so a restrictive law was not needed. The judges said that the Internet deserved the full protection of the First Amendment and noted that it broadened the public's ability to speak freely. "Through the use of chat rooms, any person with a phone line can become a town crier with a voice that resonates farther than it could from any soapbox,"[8] said Justice Stevens, who delivered the opinion on the Court's decision.

Additional Challenges

The *Reno v. American Civil Liberties Union* case was not the final time that child protection and free speech on the Internet clashed in court. Some still wanted to put Internet regulations in place to keep children from seeing sexual images online. After the Communications Decency Act received an unfavorable court ruling, the Child Online Protection Act was passed by Congress in 1998. "This has been needed for years," said Paul McGeady, general counsel for Morality in Media, an organization that seeks to maintain decency standards in the media and combat pornography. "Children will be protected from commercial smut on the Internet, and there's a lot of it."[9]

Many attempts have been made to establish online child protection laws, but none have yet been established that will protect children while not infringing on people's constitutional rights.

The Child Online Protection Act required all businesses that sold material that could be harmful to minors to restrict their access to this material. It was aimed at keeping children from having access to sexual images or nudity online. Federal law-makers wanted to make these sites inaccessible to children and argued that filtering software that parents had was not enough to protect children. They wanted the people who ran these Internet sites to make sure that children could not view them, perhaps by requiring people who accessed their site to use a credit card number to prove they were over age eighteen.

A DEMOCRATIC MEDIUM

"Ideas and controversies are erupting from every pore of American society—from blogs, talk radio, internet news and chat sites, and online video forums. The rich no longer have a monopoly on distributing ideas and views; everyone can do it, and millions are. Technology has democratized the media."—Tony Snow, former While House press secretary.

Tony Snow, "The Threat from Within," *Vital Speeches of the Day*, December 2007, p. 552.

Federal courts ruled that this law infringed on people's First Amendment rights. It was too broad and vague, and it jeopardized a person's right to receive information anonymously. The law died in 2009 when the U.S. Supreme Court refused to hear an appeal of the case. The Supreme Court's refusal to hear the case was seen as a victory for free-speech advocates. "Everyone can agree on the need to protect children from sexually explicit online material, but this misguided law tried to do it in ways that infringed too much on constitutionally protected free speech,"[10] the *New York Times* said in an editorial after the Supreme Court made its decision.

Section 230

U.S. courts twice struck down laws that would have restricted a person's free-speech rights on the Internet. While the Child On-

Newspaper Web pages are protected from lawsuits for what others post on their sites in blogs or comments areas.

line Protection Act and most of the Communications Decency Act of 1996 were ruled unconstitutional, the Supreme Court did find that one part of the Communications Decency Act actually supported free speech. It upheld a law that protects those who give others access to the Internet and allows people to post information.

This law is called Section 230. It plays an important role in giving legal protection to Internet service providers. Internet service providers give others access to the Internet and provide people with the opportunity to post information online. An Internet service provider can be a company such as Verizon or AT&T that offers customers the ability to access the Internet for a monthly fee and has telecommunications equipment that allows them to do this. It could also be a Web site host, such as Yahoo, which allows people to build their own Web site, or a social networking site like Facebook or MySpace that allows users to build a profile and post it online. A person with a blog where others can post comments is also considered an Internet service provider, as is a local newspaper that lets people post comments about news stories or local events.

Section 230 allows a broad array of Web hosts to be protected from lawsuits for what others say on their sites. Under the law, an Internet service provider cannot be sued for something posted by a person using the site. This takes the burden of control away from the service provider and means Internet service providers do not have to police their sites for defamatory content. "The intent of Section 230 was to enhance free speech and access to the Internet by protecting providers from acts that they have little or no control over,"[11] explained George H. Pike, director of the Barco Law Library and an assistant professor of law at the University of Pittsburgh School of Law.

CALL FOR MORE OVERSIGHT

"The Communications Decency Act (CDA) has generally served society well, on balance; it has been smart not to chill innovative and free speech by imposing too much liability on companies in the nascent online space. But that's not to say that all Internet intermediaries should be completely free of responsibility for anything that happens on their networks. Internet safety is one zone where greater liability might be reasonably imposed."— John Palfrey and Urs Gasser, authors of Born Digital.

John Palfrey and Urs Gasser, Born Digital. New York: Basic Books, 2008, p. 106.

The impact of Section 230 gave commentators on the Internet a more open platform than they might find in the print version of their local newspaper. A newspaper is responsible for what it prints, even when those comments come from readers, and if an inflammatory comment turns out to be false and damaging, the newspaper can be sued. However, the newspaper would be protected by Section 230 if the comment were posted on its Web site. "What a difference a legal rule can make," Tien said. "It's freewheeling. You definitely see a much less filtered view of the audience out there than you would ever get from looking at letters to the editor in any local paper. It really made it possible for a lot of free speech to happen."[12]

Because Section 230 makes the writer responsible for what is said, not the Internet provider, it takes the burden off the Internet provider for deciding what is defamatory or libelous content. "That gives a lot of Web sites and portals and intermediaries a kind [of] a safe harbor where people can speak," Tien said. "If they have to worry every time someone says something that might be defamatory or even just false and it might get them into trouble, what could they do?"[13]

Protection of Internet Service Providers

Some argue that giving a high level of protection to Internet service providers puts children in danger. In the book *Born Digital*, authors John Palfrey and Urs Gasser point to several cases in which families sued MySpace, which allows users to share photos and comments about themselves through an online profile. The lawsuits were brought against MySpace because the parents claimed the site was not protecting their children. For example, one fifteen-year-old who posted information about herself was contacted by an older man who convinced her to meet him face-to-face. He drugged and sexually assaulted her and was subsequently charged and imprisoned. The girl's parents sued MySpace for negligence in allowing their daughter to share information with the older man, but the courts said MySpace was protected by Section 230. Palfrey and Gasser disagreed with the court's decision. Although MySpace may have been protected by the law because it was an Internet service provider, they did not believe the law should prevent a parent from suing MySpace for not protecting its users' safety.

Courts have consistently ruled, however, that Internet service providers are protected. Even before Section 230 became law, courts found that Internet service providers were not responsible for what users posted on their sites. In the 1990 case *Cubby v. CompuServe*, the owner of an online news site called Skuttlebut claimed that another site called Rumorville had published remarks about Skuttlebut that were false and defamatory.

Child Protection Bills

Child online safety continues to be an issue raised in Congress. A number of bills have been introduced that seek to protect children from improper Web site content. One that was debated was the Deleting Online Predators Act, which passed the House of Representatives in 2006. It required warning labels for Web sites containing sexually explicit material. Some worried that innocent sites could be targeted if the bill passed, however. "(The Act) will cast a damaging chill over a broad range of legitimate and valuable content on the Internet, undermine voluntary labeling, tagging, and rating programs, and invite constitutional challenge,"[1] said Leslie Harris, Center for Democracy and Technology's executive director. However, Senator Conrad Burns, who sponsored an amendment to the bill, said it did not infringe on First Amendment rights. "Senator Burns' first and foremost concern is protecting our children, and this amendment essentially requires 'truth in advertising,'"[2] an e-mail message from his spokesperson said. The bill did not pass in 2006 and again failed to pass when it was reintroduced in 2007.

1. Quoted in Joan Olek, "Warning: Web-Site Labeling Ahead," *School Library Journal*, September 2006, p 17.

2. Quoted in Olek, "Warning: Web-Site Labeling Ahead," p. 17.

The Skuttlebut owner sued the owner of the Rumorville site and also sued the Internet service provider CompuServe, because Rumorville was available to CompuServe users. The court dismissed the claims against CompuServe because it did not have control over what was said on Rumorville. The New York court ruled that an Internet service provider is not liable for what is said on its site. "CompuServe has no more editorial control over such a publication than does a public library, book store, or newsstand, and it would be no more feasible for CompuServe to examine every publication it carries for potentially defamatory statements than it would be for any other distributor to do so,"[14] the court said.

Section 230 does not prevent people from taking others to court and suing if they believe their reputation has been harmed, however. A person can still sue another for defamation. If a person sues the Internet service provider who gave that person access,

Individuals like political gossip columnist Matt Drudge, pictured, can be sued for libel, unlike Internet service providers, which are protected under Section 230.

however, he will find that the Internet service provider is protected by Section 230.

The protection of Internet service providers was confirmed by a federal court in a case involving an Internet gossip columnist. In 1998 Sidney Blumenthal sued both America Online and Matt Drudge because of comments Drudge made in his online publication called *The Drudge Report*. Blumenthal claimed that Drudge had defamed him by posting false information in his report, which was available to America Online users. When looking at whether America Online could be sued, the court did not consider whether Drudge's report was true or false. It dismissed the case against America Online because the Internet service provider was protected under Section 230 of the Communications Decency Act and was not liable for what was posted on its site. Blumenthal's charge against Drudge, however, was allowed to stand because people can sue others for libel if they believe defamatory information has been published about them. The case was eventually settled out of court.

Balanced Protection

A person posting information online has been granted broad First Amendment rights from the early days of the Internet's widespread use. A person is ultimately responsible for what he or she says online but, because of Section 230, has an open platform for making a comment. Internet service providers do not need to determine whether what is said on their sites is legal. As a result, the Internet has become a robust platform for vigorous commentary.

This can lead to other issues, however, as some comments push the legal limits for defamation. Americans have the right to take others to court and sue for monetary damages if they feel they have been defamed by another's comments, but sometimes it is not easy to determine who made the comment. Comments posted anonymously raise more free-speech issues and involve Internet service providers in the controversy.

WHO SAID THAT?

The Internet is an open forum for public commentary and criticism, as well as a place where people share information. While many people post profiles of themselves online, share information about what they are doing on Twitter, or host Web sites that promote their business or political views, many others make comments using an assumed name or no name at all.

The right to comment anonymously is protected by the First Amendment. A person can hand out a pamphlet or printed notice without putting his or her name on it. Likewise, a person can post a comment on a blog site anonymously or under an assumed name, such as "Baseball Fan" or "Fantastic Shopper." People may want to make a comment anonymously in order to keep others from learning their political views or what they think about their boss.

The Importance of Secrecy

Posting information anonymously does not give a person the right to make defamatory statements, but it can protect a person who fears ramifications if his or her identity is revealed. The option brings another level of free-speech security to those who comment online. It is viewed by some as a vital protection for online speakers. "Anonymity is seen as a crucial element of the right of free speech by allowing the speech to take place without fear of retaliation or ostracism,"[15] law professor George Pike pointed out in *Information Today*.

The case of Jason Pinter illustrates why some critical bloggers want to keep their identities private. Pinter lost a job at Crown Publishing because of comments he made about the book business on his blog. While he had the right to make these comments, his employer also had the right to fire him for making them. "People like to talk about their jobs and complain about their employers; that's been true for as long as there have been jobs and employers. Pinter's experience reveals that as in other industries, some employers in publishing are apprehensive about how new forms of online employee speech might threaten office harmony or business interests," said Barry Bruce, an instructor at Vanderbilt University and author of *Speechless: The Erosion of Free Expression in the American Workplace*. "Employees like Jason Pinter have little choice but to accept limits on freedom of expression as a condition of work, or move on."[16]

What an Internet Service Provider Knows

Although making an anonymous comment provides a person with a level of privacy, that does not mean he or she cannot be sued for what is said. Online statements that are clearly defamatory are not protected by a person's right to free speech even when they are made under an assumed name. Making an anonymous threat against someone is not legal, online or offline. Ruining someone's reputation by maliciously creating and spreading

Identity Crisis

Mark Glaser, who writes a blog called MediaShift as well as more in-depth, edited pieces for the PBS Web site, noticed that people perceived him differently when he described himself only as a blogger. For example, when he gave his title as a blogger, he was told he could not receive a press pass to a music conference. He also had a difficult time being included in the Google News search engine, until he pointed out that some of his material was edited and distributed through PBS. "Whether I really am a blogger, a journalist or a blogger/journalist might not matter to me, but it will matter when I am trying to get a press pass," he wrote on his MediaShift blog.

Mark Glaser, "Am I a Journalist or a Blogger?" MediaShift, March 3, 2008. www.pbs.org/mediashift/2008/03/am-i-a-journalist-or-blogger063.html.

Even if an anonymous user posts defamatory comments online, the court can ask the Internet service provider to trace the Internet protocol address to determine the person's identity.

lies can result in a defamation lawsuit even if the person who spreads the defamatory information does so under an assumed name.

The identity of a person who posts information on the Internet can be learned even when a person posts a comment online anonymously or under an assumed name. Each computer that connects to the Internet has an Internet protocol (IP) address, a group of numbers that are assigned to that device. The number indicates a person's Internet service provider and the person's general location. When a person sends an e-mail or posts a comment on a blog, the IP number is transmitted.

The Internet service provider knows the identity of the customer behind the IP number. If a person makes an anonymous comment online that someone considers defamatory, the court can ask the Internet service provider to reveal the person's identity. The Internet service provider must weigh a person's right to privacy and anonymous speech with its legal responsibility to comply with a court order.

Under what conditions an anonymous person's name must be revealed depends on an Internet service provider's policies and can be a matter that is decided in court. Information about an anonymous user's name is typically not something an Internet service provider gives up easily. The name is usually given only if the Internet service provider receives a subpoena or court order that requires the name be revealed.

Revealing the Name

When a lawsuit is brought against a person who has posted information on the Internet anonymously, the person who claims to have been defamed must prove that a compelling reason overrules the other person's need for anonymity. Stating the name of the person being sued can be damaging, so these lawsuits

While users believe they can anonymously post on sites like Twitter and MySpace without revealing their identities, Dendrite International v. John Doe helped establish the criteria a company needs to meet in order to learn the identity of an anonymous poster.

are typically brought against "John Doe" until a judge decides whether it is necessary to reveal the name of the defendant. Although technology makes it possible for Internet service providers to know who made a post or comment because of the IP address of his or her computer, certain steps must be taken before the name is disclosed.

ANONYMITY QUESTIONS

"The question of when your [Internet service provider] or web host can be required to disclose your identity comes up in the context of libel and the courts are still struggling with it."—Chris Hansen, senior staff attorney with the American Civil Liberties Union.

Chris Hansen, interview with author, August 10, 2009.

When allegations of defamation are made, the first step is to notify the accused person of the request to obtain his or her identity. This gives him or her the opportunity to defend himself or herself. The suit must include the specific statements that are allegedly defamatory, and the person bringing the suit must show that the defamation claim is strong enough to go to trial. Pike explained:

> This requires not merely alleging that the defamation occurred but showing to the court factual and legal evidence in support of each element of the defamation claim. This includes not only the fact of the statement but proof that it is a statement of fact and not opinion, that it is factually false, and that the statement caused the victim actual harm to his or her reputation.[17]

In 2000 the case of *Dendrite International v. John Doe* helped establish the criteria a company must meet in order to learn the name of an anonymous poster. Dendrite International, which made software for the pharmaceutical industry, claimed that statements critical of the company that were posted on Yahoo! message boards were defamatory and contained trade secrets.

The company was allowed to learn the identities of two of the people who made the statements, but a third was allowed to remain anonymous because Dendrite had not been harmed by the person's statements. In making its decision the court also set guidelines for revealing the identity of anonymous posters on the Internet, noting that the plaintiff must present enough evidence to win the claim before the anonymous person's identity can be revealed. The defendant's First Amendment right to anonymous free speech must be balanced against the need for the person bringing the suit to know the poster's identity.

In another case a woman calling herself "Proud Citizen" posted comments on a newspaper's blog that criticized a councilman, calling him "paranoid" and saying he had undergone "an obvious mental deterioration."[18] The councilman filed a lawsuit and asked the blogger's Internet service provider to identify her. The Delaware Supreme Court said the blogger could remain anonymous, because the statements made were not facts. The case also illustrates the difficulty of remaining anonymous, even with the court's protection. The councilman eventually learned the blogger's identity and found that the mayor's daughter had made the comments. The lawsuit was settled out of court.

Debatable Issue

While an Internet service provider typically keeps a person's name secret until a court forces the name to be revealed, courts may use different standards to determine when that name is made public. A pair of Illinois business owners sued MyWeb-Times.com, a local newspaper Web site, in order to learn the identities of several people who had posted comments there. The business owners, who wanted to open a bed-and-breakfast, said the comments defamed them and had suggested that they had bribed planning commission members. An appeals court said that the comments had been statements of opinion, so they could be posted, but the business owners appealed the ruling and the case remains unsettled.

Sometimes the court determines that a blogger's name should not be kept secret. In another case a judge ruled that an Australian model could learn the name of a blogger whom she said

Anonymity and Credibility

When a person does not sign his name to a comment or does not name the person quoted in a blog, the credibility of the story or comment can be brought into question. Comments carry more weight when they are made by a named source. Although sources may have a good reason for remaining anonymous, a news organization that uses anonymous sources dilutes its credibility, Timothy J. McNulty noted in a *Chicago Tribune* article. "Some sources may fear criticism; others face retaliation and even physical danger for speaking out," he said. "Anonymity is a shield to protect them. But anonymity is also a way to avoid responsibility for your own words."

Timothy J. McNulty, "Opinion: Anonymous Newsmakers," *Chicago Tribune*, June 27, 2008.

defamed her on a Web site. Model Liskula Cohen asked Google to reveal the name of a blogger who called the thirty-seven-year-old an "old hag" and suggested that she was sexually promiscuous. Although the blogger's lawyer said the comments were protected because they were an opinion, the judge ruled that Cohen could sue her for defamation and that Google needed to reveal her name. Cohen's lawyer said he hoped the court's decision would show that the Internet is "not a free-for-all."[19] The blogger, Rosemary Port, in turn said she would be suing Google for failing to protect her expectation of anonymity. "Without warning I was put on a silver platter for the press to attack me," she says. "I would think that a multi-billion dollar conglomerate would protect the rights of all its users."[20]

Are Bloggers Journalists?

When a person makes an anonymous comment online, it is up to the Internet service provider to decide whether to protect that anonymity. Another issue that arises is whether a person who quotes others anonymously in an online story has the right to keep that identity private. Whether those who post information on the Internet have a right to keep the source of their information a secret is still in question.

Journalists have traditionally fought for the right to keep their sources private, and in some states they are protected by shield laws that allow them not to reveal the name of their sources. However, the far-reaching nature of the Internet allows anyone to report on the news, not only people who have been trained as journalists. What level of protection citizen journalists are allowed has been debated in court.

Print journalists are guaranteed certain rights under the law, but the extent of protection that Internet bloggers should be allowed is debated.

The right of amateur journalists to keep their sources private was questioned by Apple Computer in 2006. The company took Jason O'Grady, publisher of an online news magazine, to court after O'Grady published information about a new Apple product; Apple contended that the information was confidential. Apple asked that O'Grady reveal the source of his information, but the court ruled that he operated a news-oriented Web site and could keep his source confidential.

THE RIGHT TO REMAIN UNKNOWN

"We are concerned that setting the standard too low will chill potential posters from exercising their First Amendment right to speak anonymously. The possibility of losing anonymity in a future lawsuit could intimidate anonymous posters into self-censoring their comments or simply not commenting at all."— Judge Myron Steel.

Quoted in Steve Yahn, "Courting Blogs," *Editor & Publisher*, August 2006, p. 24.

Another amateur journalist's rights were also called into question when he filmed a disturbance in 2005. Josh Wolf was jailed after posting a video of an anti-capitalist rally in San Francisco. At the rally a police officer was severely injured and a police car was damaged. Wolf filmed the event and posted the video on his blog. He also sold most of the footage to a local TV station.

Investigators asked him to release the rest of his footage and identify the protesters. Wolf refused and was arrested and jailed for contempt of court, but claimed that he was protected by the First Amendment. He spent several months in jail before being freed. "When the Constitution and the First Amendment were written, obviously there weren't such things as bloggers," Wolf commented. "But the pamphleteers were self-published—they were the eighteenth-century equivalent of a blogger. So the Constitution had people like me at its core."[21]

Protections and Limits

The First Amendment protects people such as Josh Wolf who seek to bring information to the public. It allows anonymous criticism and commentary, providing an open forum for discussion while guarding a person's privacy. While protections are given to people posting information, a blogger's right to remain anonymous is balanced with others' concerns over defamation. To make sure the diverse and spirited commentary on the Internet does not override people's right to preserve their reputation, people can take legal action if they believe they have been harmed by what is said or shown.

In addition to libel concerns, other free-speech issues arise online, however. Josh Wolf was clearly the owner of the videotape he took and could post it where he pleased. But what about others who use the videos or commentary of others to help them make their own points? Wolf had his own Web site for posting the video and could put it up or take it down at will. What about others who do not have the luxury of Web site ownership? Who owns the content posted on the Web and who should be able to take information down are further issues relating to online free speech.

CHAPTER 4

COPYRIGHT AND CRITICISM

Technology has made copying and posting text, photos, and videos onto a Web site as easy as typing a few buttons on a computer keyboard. While technology has made it easier to copy information, the ownership of that material remains subject to laws that have been in place for hundreds of years. Copyright laws give the creator of a work, such as a songwriter, the author of a book or play, or a video's creator, control over that work. The creator of a literary or artistic work has the right to make copies of the work, publish, and sell it for a period of time under copyright laws.

Copyright laws protect a person's right to be the exclusive rights holder for a work. They guard a person's ability to be paid for copies of his or her work and make sure the work is not changed without permission. These laws have changed over the years to include works such as records and videos and reflect changes in technology such as the photocopier and computer, which have increased the speed with which copies can be made. Copyright laws continue to perform the necessary function of controlling the rights to copy a work while recognizing the public's interest in owning a copy of that work or build on its ideas. "Copyright law is our way of balancing the needs of creators (to make a living, to preserve their work's artistic integrity) against the public's justifiable desire for knowledge and works of art,"[22] explained Chuck Leddy in *Writer* magazine.

Copyright Act

Under copyright law, a person owns the rights to his or her work for a period of time. For works created in or after 1978, this period of time is the creator's lifetime plus seventy years. A person's descendants can renew the copyright for another seventy years. After the copyright on a work expires, the work goes into the public domain and can be used by anyone.

Older works were also governed by copyright law that protected them for a period of time. The copyrights of many older works have expired, placing the works in the public domain. The works of Shakespeare, Mary Shelley's *Frankenstein*, and Mark Twain's *The Adventures of Huckleberry Finn* are all in the public domain and can be duplicated or used by anyone if credit is given to the original author.

In protecting a creator's work for a period of time, copyright laws are designed to prevent copyright infringement or piracy, the unlawful copying of a work. The laws that govern copyright were updated in the late 1990s to keep up with Internet technology. The ease with which information can be shared on the Internet made modification necessary, and in 1998 the Digital Millennium Copyright Act was signed into law by President Bill Clinton. The bill regulates the use of copyrighted material on the Internet, making it a crime to get around the anti-piracy

Copyright laws protect the rights of an author and his or her work and are intended to prevent piracy.

measures that prevent copyright infringement. It makes it illegal to sell, make, or distribute code-cracking devices that are used to copy software illegally; a person cannot provide computer code to others that helps them get around the software that protects copyrighted material. The act impacted computer software, video games, songs, and music as well as books. Web sites that made it possible to download songs or computer software for free were subject to copyright infringement lawsuits.

Fair Use

While the act was designed to update copyright law to keep up with technology, some object and contend that the anti-piracy provisions of the copyright law go too far. It is not questioned that copying and selling someone else's work is illegal, but a concept called fair use allows people to use a portion of a copyrighted work for instructive purposes. A researcher can quote

An Inventor, the Internet, and Free Speech

A man whose dance was shown in an online video used copyright law to question the right of others to use material showing the dance. In a case involving the electric slide, Richard Silver, who claims to have created the dance, attempted to make video creator Kyle Machulis take a concert video off YouTube because his video included a ten-second segment of the audience doing the electric slide. Silver eventually reached a settlement that licensed the dance under a Creative Commons license, which allowed it to be performed for noncommercial purposes, but his use of copyright law to ask that the video be removed from the Internet was questioned. "Silver's claim of copyright infringement is absurd and is a classic example of the kind of [Digital Millennium Copyright Act] abuse that can chill Internet speech," said Corynne McSherry, a staff attorney for the Electronic Frontier Foundation. "Even if Silver had a valid copyright in the dance—which is not at all clear— this is a fair use and not infringing."

Quoted in Electronic Frontier Foundation, "Electric Slide Creator Steps on Fair Use," March 1, 2007. www.eff.org/press/archives/2007/03/01.

academic works in his or her papers and a news reporter can quote part of a letter or copyrighted book without breaking the law. Fair use allows an author to protect his or her rights to the work but also allows the public to use a portion of the work to criticize, comment, and otherwise engage in free speech. However, the restrictions on computer code in the Digital Millennium Copyright Act raised concerns that electronic locks on digital material would make it difficult for a person to gain access to the work in order to make fair use of it.

The issue went to court in 2000. Several people had been distributing a computer code that allowed people to get around the encryption software on a DVD. A number of movie studios feared the code would allow illegal copies of movies to be made. The studios filed a suit against a man who posted the software on his Web site and linked to other sites that also contained the software. The defendants argued that the First Amendment allowed them to share the code in order to give people the opportunity to make fair use of copyrighted works, but a district court ruled that sharing the code was in violation of copyright law, which restricts descrambling programming. The decision was appealed, but an appeals court upheld the lower court's decision. The courts found that the Digital Millennium Copyright Act was constitutional and did not restrict a person's right to free speech.

Book Project

While the courts decided that the limits on anti-piracy software could stand, concerns were also raised that other technology continued to make it too easy to copy and have digital access to published works. Internet giant Google began scanning and digitizing millions of books in 2002. The project allows users to read books online, download books that are in the public domain, and use its search engine to search through text for information. Because its catalog includes books that are still under copyright, the project raised concerns of the Author's Guild and the Association of American Publishers. Google had not asked for permission to scan the copyrighted books, and the guild and publishers' association filed a lawsuit against Google.

Converting books to digital format provides a vast amount of easily accessible knowledge.

Google claimed that it was allowed to post portions of copyrighted text online under fair use rules. Only a portion of the copyrighted works was visible online, and Google contended that was in line with what fair use allowed. This still concerned writers, however, who worried about being compensated for their work. "What's at stake in these copyright battles is the future of our literary culture. If authors can't pay their rent, what incentive will they have for creating literary works?" asked Leddy. "While authors write to express their artistic vision and make their voices public, they also have a right to protect their work and profit from it."[23]

Google reached a settlement with the authors' groups in 2005 for $125 million that involved a royalty system for authors and publishers. The case continues to wind its way through the court system, however, as the settlement must be approved by a federal judge. Google's actions are copyright infringement, according to Marybeth Peters, the register of copyrights at the U.S. Copyright Office, and allowing Google to reach a settlement regarding digitized books would give Google new rights and take away Congress's ability to change copyright law. Peters said:

> The settlement would alter the landscape of copyright law, for millions and millions of rights holders of out-of-print books. It would flip copyright on its head by allowing Google to engage in extensive new uses without the consent of the copyright owner—in my view, making a mockery of Article One of the Constitution that anticipates that authors shall be granted exclusive rights.[24]

When the U.S. Department of Justice commented on the settlement, it expressed concerns but also noted the value in allowing access to digitized books. Questions arose over the right to works that were out of print or for which the copyright owner could not be found, and the impact the agreement would make on competitors who also wanted to digitize books. While the Department of Justice asked a judge to reject the settlement, it also urged that negotiations continue. It recognized that having books easily accessible in digital format could help researchers and others who wanted to locate books that were difficult to find and that the project could be beneficial to society. The Center for Democracy and Technology called the project "extraordinarily valuable" and said it "will make available to the public a vast amount of knowledge and information that is largely inaccessible today."[25]

The Right to Criticize

Having access to portions of text or video in order to make a point in a commentary is part of a person's First Amendment right to free speech. Illustrations can be used to make a point as well, and courts have held that fair-use law also applies in that respect. The issue was brought up in court when a former employee with a gripe against his old company used technology to express his opinion about his ex-bosses. He copied photos of company executives from the Internet and mailed out postcards that had the photos morph into images of Adolph Hitler and Heinrich Himmler. The company said the photos of company executives could not be used because of copyright.

A NEED FOR CONTROL

"If they are to thrive, authors, moviemakers, painters, software creators, and others do need a way to control commercial uses of their work. Preserving copyright looks to be the best way to achieve this goal."—Pamela Samuelson, professor of information management and law at the University of California–Berkeley.

Pamela Samuelson, "The Digital Rights War," *Wilson Quarterly*, Autumn 2008, p. 48.

Copyright law's fair use provisions allow snippets of information created by other people to be used to emphasize a point in an entirely new creation. In this case the photos the employee used were not his, but he used them to make a powerful point that was. His right to use the photos as part of his criticism was held up in court. The disgruntled employee who used photos of his bosses was able to use the pictures, the court ruled, because the photos were integral to his criticism and were not used for commercial purposes. The photos fell under the fair use provision of the copyright code that allows a copyrighted work to be used for criticism and comment.

Improper Copyright Use

Copyright law is sometimes used by companies in an attempt to stop people from criticizing one of their products. The Digital Millennium Copyright Act received its first challenge in this respect when information about flaws in voting machines was posted online. The information came from internal e-mails at Diebold, which makes voting machines. Diebold said that the Internet sites that posted these documents infringed on its copyright and demanded that the sites take the information down. A company can ask that an Internet service provider remove information from its Web site that is in violation of copyright law, but it cannot use copyright law to try to get information removed from the Web if a copyright infringement has not occurred. In this case the court ruled that the discussions were not an infringement of copyright and that the company knew it was not. "No reasonable copyright holder could have believed that the portions of the email archive discussing possible technical problems with Diebold's voting machines were protected by copyright,"[26] Judge Jeremy Fogel said. Under copyright law a person or company that tries to use the law to get information removed from the Web when it knows an infringement has not occurred must pay damages, and in this case Diebold paid more than one hundred thousand dollars in damages, including court costs and attorney fees.

Video and Fair Use

Copyright law has also been used to try to stop video clips from being used as part of criticism. In May 2009 blogger Perez Hilton asked YouTube to take down a video posted by the National Organization for Marriage. The video contained several seconds

Expensive Lawsuit

Breaking copyright law by sharing music online can be an expensive proposition. A Minnesota jury found Jammie Thomas-Rasset guilty of copyright infringement and ordered her to pay $1.9 million in July 2009. She was initially accused of illegally sharing more than 1,700 songs. The Recording Industry Association of America, which brought a lawsuit against her, reduced the number to a representative sample of 24. The jury ordered her to pay $80,000 for each song.

The thirty-two-year-old woman said she would appeal the decision, attorney Joe Sibley said. "She wants to take the issue up on appeal on the constitutionality of the damages," Sibley said. "That's one of the main arguments—that the damages are disproportionate to any actual harm."

Quoted in Greg Sandoval, "Jammie Thomas Will Appeal, Lawyer Says," CNET News, July 1, 2009. http://news .cnet.com/8301-1023_3-10277701-93.html.

Video sites such as YouTube have received attention and caused debate regarding copyright laws.

of a clip of Hilton, showing him strongly criticizing Miss California Carrie Prejean. Hilton claimed that the video violated his copyright to the material and so it should be removed. The organization countered that it had the right to use the clip of Hilton to make the point that the organization's opinions were often harshly criticized.

YouTube quickly recognized that Hilton did not have a valid copyright argument and agreed with the National Organization for Marriage. It had removed the video from its site at Hilton's request, but soon put the video back online. The organization could include Hilton's clip in its video because it was using the clip to express its opinion. Just as Hilton had the right to criticize Miss California, the National Organization for Marriage had the right to use Hilton's words to make its own argument. Expressing his own opinion on the issue, Tim Jones, a blogger for the Electronic Frontier Foundation, called Hilton's request to have the video taken down a "baseless attack on free speech," noting that the National Organization for Marriage's "use of Hilton's video clip was clearly fair and non-infringing—it is brief, transformative, critical and does not pose a competitive threat to Hilton's market."[27]

Other video clips have also been found to be fair use. Use of a video that was critical of Uri Geller, who claims to be able to bend spoons with his mind, was debated in a 2007 copyright case. Geller sued Brian Sapient, who had posted videos on YouTube that tried to debunk irrational claims. One of his videos included an eight-second clip of Geller, which had been uploaded from a *NOVA* program called "Secrets of the Psychics." Geller demanded that the video be taken down because it infringed on copyright, and YouTube took down his videos. Sapient, backed by the Electronic Frontier Foundation, said that the clip was fair use under copyright law. A settlement was reached in the lawsuit, which allowed for fair use of the material.

Trademark Violation

Trademark law is another area that can make an impact on a person's right to free speech. While copyright law protects original creative works, trademarks protect a company's brand, or sign, under which its products are distributed. Company names, logos, and phrases can all be trademarked, and a company's right to protect its trademark can be at odds with another's right to criticize that company. A person posting information or commenting on the Internet may argue that use of the trademark is protected under the First Amendment because it is part of a message, but the company may argue that using a trademarked name, term, or logo causes confusion about the source of goods or services.

The right to post information critical of a company or organization by making fun of its name was questioned when a group of bicycle enthusiasts staged a demonstration at the Chicago Auto Show. They also posted a parody Web site, Auto Show Shutdown. They wanted to alert the public to the dangers of global warming and problems caused by cars. The Chicago Automobile Trade Association said the site violated the organization's trademark, and the association wanted the bicycle group to take down the site. However, the bikers said they had the right to post the site under the First Amendment, and the issue was dropped.

Parody Problems

Making a point through parody, which exaggerates certain aspects of a subject to make a point or for the purpose of humor, can lead to confusion if it is not obvious that the site has nothing to do with the makers of the original material. The *New York Times* contacted the Republican Governors Association because its satirical Web site looked too much like NYTimes.com. The site violated copyright laws, the *Times* said, and caused confusion by suggesting that the *Times* was somehow linked with the site. In response, the association made some changes to the look of a site called Corzine Times, which targeted Jon Corzine, a Democratic governor from New Jersey.

PROVIDING PROTECTION

"Through enactment of the Digital Millennium Copyright Act, we have done our best to protect from digital piracy the copyright industries that comprise the leading export of the United States."—President Bill Clinton on signing the Digital Millennium Copyright Act on October 28, 1998.

President Bill Clinton, "Statement on Signing the Digital Millennium Copyright Act," *Weekly Compilation of Presidential Documents*, November 2, 1998, p. 2168.

A New York blogger had to change the domain name of her Web site and add a disclaimer to her site after her commentary about a redevelopment project in the city raised concerns. The woman made a parody Web site targeting a project at Union Square in New York City, but the site was shut down for six months because of concerns that her site caused too much confusion with the project's real Web site. The Union Square Partnership brought a copyright lawsuit and other legal action against the Web site's creator, Savitri Durkee. After the case was settled out of court, Durkee relaunched her site. She still felt strongly that the point of view she had to make should be available. "It's important that we have uncensored debate about what comes next for this historic landmark,"[28] Durkee said.

The site remained an important example of how the Internet could be used to make a political point, noted Corynne McSherry, a staff attorney for the Electronic Frontier Foundation, which represented Durkee. "The Internet has revolutionized political activism, allowing everyone to have his or her say on a global platform," McSherry said. "Parody is an important part of America's free speech tradition, and with her online parody Ms. Durkee brings this protected form of criticism into our digital age."[29]

Used within the law, parodies of the works of others or commentaries that use quotes or photos to make a point add a strong component to Internet free speech. However, the person who posts the information also takes responsibility for making sure that his or her site does not take away from the information it uses but is rather a new creative work. He or she also bears responsibility for making sure posts do not violate a company's trademark. These levels of protection allow free speech to happen but also give others a course of action to take if they believe they have been damaged by what has been said.

SCHOOL SPEECH AND CYBERBULLIES

Students have special protections and responsibilities when it comes to Internet free speech. Parents have the responsibility to oversee or filter what their children access online. Students have restrictions about what they can and cannot say about their school online, as well as restrictions when it comes to bullying others online, called cyberbullying. While online communication, Facebook sites, and MySpace pages are part of many students' everyday lives, they are also sources of debatable free-speech issues.

Free Speech for Students

Students typically are quick to embrace new technology, and many are avid users of social networking sites and online communication. As students publish information on sites such as YouTube, Facebook, and Twitter, schools find themselves with new questions over what should be regulated and how to regulate it. Writer Carol Brydolf noted in *Education Digest*:

> Administrators and school boards across the country are struggling to lawfully manage their students' use of social networking sites like MySpace, Facebook, LiveJournal, and Xanga. These sites make it easy for students to post photos, personal information, video clips, and music files and to build networks of "friends" across the country, many of whom they may never have met in person. More and more, schools find themselves caught between their legal and moral obligation to provide a safe environment that promotes learning and their students' constitutional rights to free expression and privacy.[30]

The rights of schools to discipline students for what they say online is balanced with a student's right to express an opinion. Also taken into account is the impact student comments make on others. Debate arises over how much oversight and restriction is needed for keeping order at school and whether it compromises a student's right to free speech.

Decades before the Internet came into widespread use, the U.S. Supreme Court decided that students do have constitutional rights in school. In the 1969 case *Tinker v. Des Moines Independent Community School District*, students wore black armbands to school to protest the war in Vietnam. The court determined that the passive protest was an allowable expression of their views under the First Amendment.

Mary Beth Tinker and her brother, John, both pictured, are the namesakes for Tinker v. Des Moines Independent Community School District. *In 1969, the Supreme Court determined that students have constitutional rights at school. The development of online communication, however, has caused schools to question how to regulate the information students publish online.*

The Internet has stretched school discipline boundaries. Students can receive punishment at school for harmful or inappropriate comments they post even off of school property.

On and Off Campus

While *Tinker* made it clear that students have rights at school, the Internet has stretched the boundaries for discipline by school authorities. The Supreme Court has not yet ruled on a case involving student free speech online, but cases have come before lower courts. In light of lower court decisions, students may find themselves disciplined at school for comments made on a MySpace page or in an e-mail they sent to friends from home—and they may find themselves unprotected by the First Amendment. "Students do need to be very careful of what they say," says Mary-Rose Papandrea, a member of the Boston College Law faculty. "The law is unsettled. Right now it's pretty dangerous for kids to post stuff, even if it is just a private e-mail to a friend. If it gets into the hands of school officials it's anyone's game right now."[31]

A New York eighth-grade student who used a violent image as his IM icon found that his school could discipline him for what he created at home. The icon, a small drawing that appeared next to his name when Aaron Wisniewski sent instant messages to his friends online, portrayed a gun pointing at a head, with the word "kill" and his English teacher's name underneath the image. After the icon came to the attention of school officials, he was suspended for a semester for breaking school rules, threatening a teacher, and disruption. His parents sued the school, claiming that the icon was not a true threat and should be allowed as free speech, but a judge said that the discipline was allowed. "The fact that Aaron's creation and transmission of the [instant messaging] icon occurred away from school property does not necessarily insulate him from school discipline," Judge Jon Newman said. "We have recognized that off-campus conduct can create a foreseeable risk of substantial disruption within the school from some remote locale."[32]

SCHOOL SPEECH LIMITS

"You can express an opinion on whether someone is a good teacher. But when you start inviting people to say that they hate a teacher, that crosses the line. . . . We don't want teachers to work in fear, looking over their shoulders when they walk to their cars after school."—Pamela Brown, assistant director in the Broward County School District in Florida.

Quoted in Carmen Gentile, "Student Fights Record of 'Cyberbullying,'" *New York Times*, February 7, 2009. www.nytimes.com/2009/02/08/us/08cyberbully.html.

Disturbing Comments

Speech does not have to be violent in order for a student to get into trouble. A Mississippi cheerleader who had a profanity-laced conversation on her Facebook site with another cheerleader was banned from practices, games, and events after her cheerleading

Online Policy

One option for schools in dealing with Internet free-speech issues is to implement a policy that addresses how the school will deal with students' rights to free expression. The Ball State University J-Ideas Web site, which supports high school journalism training, suggests that such a policy include a commitment to education about free speech, such as information on censorship, online safeguards, media literacy, and a parent's role in the issue. The abridged version of the policy recognizes both student free-speech rights and responsibilities, saying:

Student-produced media are sponsored by the school and must comply with established rules forbidding material that is libelous, obscene or materially disruptive to the educational mission. The school board recognizes that students have a constitutional right to free speech within the limits of those rules and will foster an atmosphere in which responsible free expression is taught and encouraged.

J-Ideas, "Student Expression Policy." www.jideas.org/abridged_digital_policy.php.

instructor saw it. The student filed a lawsuit saying that her free-speech rights were violated, noting that she thought the private messages were not something the teacher should have accessed. Her lawyer said the teacher's actions crossed the line. "It's egregious to me that a [then] 14-year-old girl is essentially told you can't speak your mind, can't publish anything, can't be honest or have an open discussion with someone without someone else essentially eavesdropping,"[33] said her attorney, Rita Nahlik Silin.

In another student free-speech case, a U.S. Court of Appeals ruled that school officials could discipline Avery Doninger, who wrote in her blog that "jamfest is cancelled due to douchebags in central office," and said those who read the blog should get in touch with the superintendent "to piss her off more."[34] School officials took away her ability to serve as class secretary or speak at her graduation, and the court said the school did have that power. Some said the case had a chilling impact on students' right to free speech. "This demonstrates how narrow students'

First Amendment protections are," said Mitchell H. Rubinstein, an adjunct law professor at St. John's University and New York Law School. "If this student can't speak up against the administration, what student can?"[35]

The Wisniewski and Doninger cases illustrate how what a student says at home or to friends is not always a private matter. Critical comments that have a tie to school could get a student in trouble. "It used to be easy to know if kids were speaking in school or out of school, now the boundary is not that clear," Papandrea said. "Students have always made fun of their teachers. Now the broad community may see something the student thought was just for him and his buddies. I think a lot of kids think no one else is seeing what they're doing, but that's not always true."[36]

Legal Consequences

Teens face more than school disciplinary actions when the information they post online breaks the law. Their actions can also have legal consequences. A Wisconsin teen who posted nude photographs of his sixteen-year-old former girlfriend on a MySpace page was charged with criminal libel, causing mental harm to a child, and sexual exploitation of a child. After he pleaded guilty to causing mental harm to a child, the other charges were dropped. He had to serve one hundred hours of community service and three years of probation, during which he could not own a computer or other device with Internet access capability.

Another Wisconsin teen also faced criminal prosecution after he created a MySpace page for a high school police officer that alleged that he lied, hit on underage girls, and harassed students. The student later apologized to the officer and said he did not realize the full implications of what he had done. He was charged with defaming the police officer, but authorities deferred prosecution as long as the teen stayed out of trouble. The teen was allowed to pay a fine for disorderly conduct.

Cyberbullies

The sixteen-year-old whose photos were posted on the Internet and the police officer who was ridiculed on MySpace are both victims of digital-age crimes, as the online world puts criminal activity in a new dimension. The Internet can be a source of contention and emotional distress when spiteful and malicious comments are posted. Students connecting with others online run the risk of coming into contact with inappropriate material.

The use of the Internet can lead to cyberbullying, in which users send threats and harassing comments, e-mails, or text messages to another person.

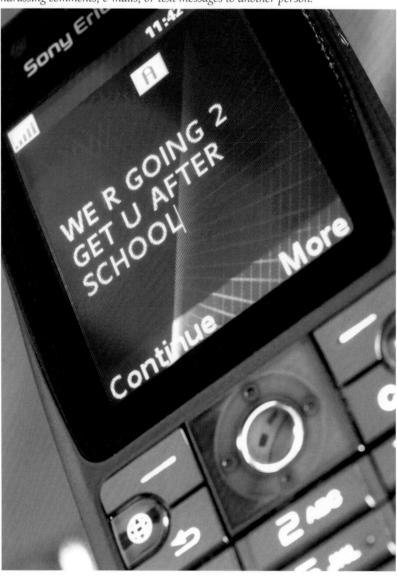

While the Internet is a popular form of communication and allows for quick and spirited conversation, Internet chats or personal Web sites can also degrade into a series of put-downs, derogatory comments, and threats. *Cyberbullying* is defined as harassing text messages, instant messages, voice mails, e-mails, and comments made in chat rooms or on social networking sites. It can bring on emotional distress and may also lead to substance abuse and depression. "The assault can come from so many directions that a youngster feels bombarded and uneasy in any environment,"[37] says Edwin C. Darden, an attorney and contributing editor to the *American School Board Journal*.

NEW BATTLEGROUND

"Students need to remember what they post online, unlike a generation ago when people wrote in diaries, what they do on their social networking MySpace or Facebook, is available for everyone to see. . . . This concept of digital free speech is really the new battleground of the First Amendment, and we must determine where the line is drawn in cyberspace."—Warren Watson, director of J-Ideas, a First Amendment institute at Ball State University.

Quoted in Kelsey Beltramea, "Tangled Web," Student Press Law Center, Fall 2008. www.splc.org/report_detail.asp?edition=46&id=1439.

The Internet broadens the reach of those making cruel comments. A MySpace or Facebook page is an opportunity for students to share information about themselves, but it can also become a battleground filled with hateful messages. "Youngsters with a penchant for cruelty, whose assaults on classmates were once confined to a single classroom or campus, can use the Internet to broadcast their insults or threats to a wide audience," said Brydolf. "Going online to torment others, or 'cyberbullying,' has become a national problem."[38]

Internet Bullying 2006

Have received mean or threatening emails

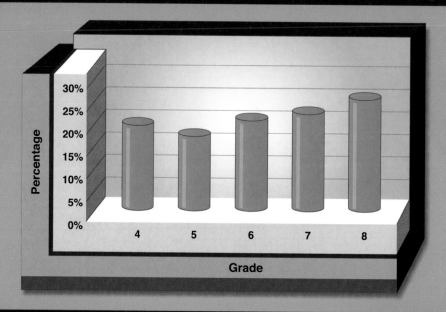

I have e-mailed hurtful or angry things to another person

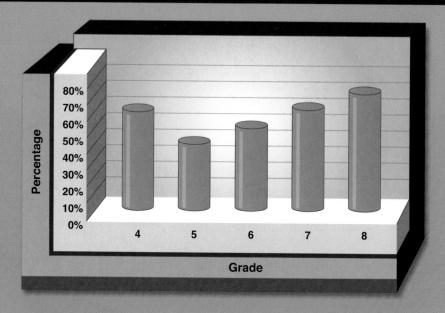

Taken from: isafe.org, http://www.isafe.org/channels/sub.php?ch=op&sub_id=media_cyber_bullying.

Legal Implications for Cyberbullies

Whether mean-spirited, degrading, humiliating, or nasty comments are simply an extension of free speech or a basis for a defamation claim is an issue that has been brought up both in court and in Congress. Those who find their lives impacted by the online commentary of others may take their case to court, while Congress considers whether additional rules and oversight are needed to try to curb harassment.

A New Jersey teen was so hurt by comments others made about her on a Facebook page that she took her case to court and sued four classmates for $3 million. She also sued Facebook over the page that questioned her morals and that asserted she had AIDS and used drugs. "She had a very difficult time in high school," her attorney, Mark Altschul said. "They were making sure she was an unwanted soul there."[39] The suit was filed in the spring of 2009 and is still pending.

SOCIAL NETWORKING SPILLOVER

"Most of the time, social networking sites provide adolescents and young adults viable and healthy methods of self-expression. However, when those sites are used to propagate disruptive, slanderous, or even threatening speech, educational environments are often impacted."—Steven M. Baule, school superintendent in Westmont, Illinois, and Darcy L. Kriha, Chicago lawyer.

Steven M. Baule and Darcy L. Kriha, "Free Speech in a MySpace World," *Library Media Connection*, February 2008, p. 22.

Students are not the only ones impacted by comments from cyberbullies, as teachers and other school administrators can also find themselves the target of hateful online comments. A Florida high school teacher was criticized on the Facebook page of high school senior Katherine Evans, who said the woman was the worst teacher Evans had ever had. Evans also invited other students to express feelings of hatred for the teacher on her blog.

The post was removed, but Evans was still disciplined; she was suspended from school and removed from her Advanced Placement classes. She sued the school principal, saying her First Amendment rights were violated. Her attorney said the case is an example of the questions raised by school discipline for off-campus speech. He pointed out that what she said did not threaten anyone with harm. "She has the absolute First Amendment right to do this," lawyer Matthew Bavaro said. "The question is how far does the school's authority go to punish off-campus speech they

John Halligan shows the Web page dedicated to his son Ryan who killed himself after months of cyberbullying.

don't like? If Katie had praised the teacher, would she have been punished? The school is judging what is appropriate speech."[40]

Tragic Implications

The consequences of cyberbullying can go beyond a damaged reputation and school suspension, however. At its worst, on-line harassment has a devastating result. Teen Megan Meier of Dardenne Prairie, Missouri, killed herself after she received painful messages on MySpace from a person she thought was a boy she had befriended online. The person was really Lori Drew, the mother of another girl who knew Megan and had had a falling out with her. Drew had created a fake MySpace profile and pretended to be a boy. Megan killed herself after receiving a message from Drew that said, "The world would be a better place without you in it."[41]

Limits of the Law

"The law shouldn't stand in the way of parents bringing a claim of negligence against an online provider when a child has been harmed. There is no reason why a social network should be protected from liability related to the safety of young people simply because its business operates online."—John Palfrey and Urs Gasser, authors of *Born Digital*.

John Palfrey and Urs Gasser, *Born Digital*. New York: Basic Books, 2008, pp. 106–107.

Drew was charged under the Computer Fraud and Abuse Act for using false information to set up her MySpace account for the purpose of harassing or harming Megan and was also charged with intending to inflict emotional distress. Drew was convicted of violating the MySpace terms of service, but the jury did not find sufficient evidence to convict her of intending to inflict emotional distress. A judge in a higher court later decided that Drew had not violated the Computer Fraud and Abuse Act with her actions, either. In July 2009 that conviction was overturned on the grounds that it was unclear that she

Online Speech Education

Mary-Rose Papandrea, a member of the Boston College Law School faculty, questioned both the legality and wisdom of allowing schools to impose restrictions on student online speech. In a blog commentary on the online Citizen Media Law Project, she noted that while teachers have the right to maintain order the in the classroom, it is not as clear whether this extends to what students say and do online. "Clearly it is unobjectionable for [a] teacher leading a physics lesson to tell students that they can't talk about the presidential election during class," she said. "For the most part, however, communications on the Internet do not intrude into the public space." She added that a better course of action would be to teach students proper online commentary. "Without some education about how to exercise their free speech rights, students would enter the adult world without the necessary skills to contribute to the political world," she said, concluding that, "for the most part, the primary approach that schools should take is not to punish their students for their speech on the Internet, but to educate them about how to use this medium responsibly."

Mary-Rose Papandrea, "Schools Lack Authority to Punish Online Student Speech," Citizen Media Law Project, November 13, 2008. www.citmedialaw.org/blog/2008/schools-lack-authority-punish-online-student-speech.

broke a federal law when she violated the terms of service for MySpace.

While the idea that a mother would cause such pain to a child horrified many, the fact remained that Drew's actions were not illegal at the time. Andrew Grossman, a senior legal analyst with the Heritage Foundation, a conservative public policy research institute, explains: "What happened to Megan is truly a tragedy and no one wishes to downplay that. Ms. Drew didn't do anything that was against the law. She did some things that were unkind, that were rude and not becoming of an adult but not against the law."[42]

Stopping Cyberbullies or Limiting Free Speech?

In response to the Megan Meier tragedy, Missouri passed legislation in 2008 making cyberbullying a criminal offense, with penalties ranging from a year to five years in prison. A national law to prevent cyberbullying, the Megan Meier Cyberbullying Prevention Act, was introduced in Congress in 2009. California congresswoman Linda Sanchez commented:

> We need to make new laws in response to these new crimes. . . . What they need to know is that cyber bullying is a serious crime, and is no less harmful than in-person threats, stalking, and harassment. If federal law recognizes this new form of bullying, police and prosecutors will be better equipped and educated to deal with this problem. Prosecutors, more importantly, will then have the ability to punish this behavior in court.[43]

Some questioned whether the legislation was necessary, however, and wondered if it infringed on free speech. John Palfrey of the Berkman Center for Internet and Society at Harvard University said:

> One of the big questions we have to grapple with is whether or not bullying done online makes us feel any differently than bullying in the old fashioned way. If it doesn't then it's not clear we need a new law in this context. Generally speaking it's a bad idea to make cyber specific laws. We don't need a cyber law for stealing, we have a law for theft. We don't need a cyber law for fraud, we have basic laws for that. And the question now should be do we want to outlaw bullying and not just cyber bullying?[44]

The question over whether a law against cyberbullying would help stop harassment or infringe on free speech is an issue that has yet to be decided. Lawmakers, educators, and students must contend with questions over balancing a person's right to be safe and the right to speak freely. The Internet makes an impact on the social lives of people and can take its toll on their emotional state as well. Regardless of the regulations on the issue, it is clear that comments made in cyberspace can have far-reaching consequences.

INTERNATIONAL FREE SPEECH

While the Internet has been a protected forum in the United States, that is not the case in every nation. Oppressive governments such as those in China, Iran, and Burma (Myanmar) restrict freedom of expression and impose controls over what their citizens read and write. Sharing information in countries where free speech is restricted can be dangerous, as repressive governments go after this if they fear such speech is subverting law and order. Those who make statements that disagree with the government's stand risk arrest and imprisonment.

The Internet has complicated these governments' abilities to inhibit free speech, however. Citizens take advantage of technology to share news and information that would have otherwise been censored. They also use Web sites and Twitter to make comments, although they face serious consequences if their actions come to the attention of government officials who see their remarks as dangerous.

A Free-Speech Option

Because it offers a degree of anonymity, online speech has the potential to give a person the opportunity to make a controversial statement or advance an idea that does not agree with the beliefs of those in control. Those in countries where free speech is restricted can attempt to make comments without repercussions by using filters that hide their identity, or they may have their Web sites hosted in another country. In this way the Internet can offer a citizen's view of what is taking place and foster the sharing of ideas. It also empowers people by giving them access to information.

Broken Promise

In Thailand the leader of a military coup that overthrew the government promised more free speech on the Internet. Instead, however, after Prime Minister Surayud Chulanont came to power in 2006, things became more restrictive. Human Rights Watch, an independent organization dedicated to defending human rights, accused the government of undermining free political debate by blocking Web sites and monitoring opinions that are critical of the government.

"A major complaint about (former Prime Minister Thaksin Shinawatra) was his muzzling of the media and willingness to limit free speech," said Brad Adams, Asia director of Human Rights Watch. "The military-backed government promised a quick return to democracy, but it's now attacking freedom of expression and political pluralism in ways that Thaksin never dared."

Brad Adams, "Thailand: Military-Backed Government Censors Internet," Human Rights Watch, May 22, 2007. www.hrw.org/en/news/2007/05/22/thailand-military-backed-government-censors-internet?.

Commenting online is not always a safe thing to do. People who wish to speak freely on the Internet may risk jail or death if their comments upset government officials. They are in a constant technological battle with their governments as they try to use the Internet to speak freely and their governments use technology and restrictive laws to stop them.

The free-speech stakes are high in countries where the ability to express an opinion is not a right protected by the government. In Burma (Myanmar) a blogger was jailed for distributing video footage of a hurricane. An Iranian blogger who was imprisoned for making insulting remarks about religious leaders in the country died in jail. Bloggers have also been jailed in Cuba, Saudi Arabia, and Egypt, according to the Committee to Protect Journalists, which lists those countries as among the ten worst places in the world to be a blogger.

Blocking Free Speech

In an attempt to control free speech online, some nations restrict Internet access and block information. The OpenNet Initiative, a

Some nations restrict information and Internet access in an attempt to control free speech.

partnership between four academic institutions that aims to investigate and expose Internet surveillance, estimates that twenty-six nations censor the Internet, including China, Iran, Syria, Pakistan, Tunisia, Vietnam, and Uzbekistan. Some entire services, such as YouTube, Skype, and Google Maps, are blocked. In Burma (Myanmar) few people have Internet access in their homes, so most go online in cybercafes. However, in these cybercafes the government can monitor e-mail messages and block Web sites. In Syria, Facebook and some Arab newspaper sites were among the 255 Internet sites blocked in 2008, according to the Syrian Media Centre. In the United Arab Emirates, Human Rights Watch notes that there are new media laws pending that would restrict media freedom.

China has been especially cautious when allowing its citizens Internet speech rights. Bloggers must deal with government censorship tools designed to keep anti-government opinions from being posted on Internet blogs, and the nation's leaders try to crack down on what it calls false news. China has also attempted to block references to topics such as democracy and human rights, banning subjects such as the Tiananmen Square and Tibet.

Chinese citizens who do not comply with these limits face arrest. In 2004 journalist Shi Tao was arrested and jailed on charges of subversion for sending an e-mail describing the Chinese government's efforts to downplay the fifteenth anniversary of the pro-democracy uprising in Tiananmen Square. The message was sent to a Web site based in New York that advocated democracy. China convicted him of divulging state secrets, and in April 2005 he was sentenced to ten years in prison.

Some say that even with limits, having the Internet in China brings more information into the country than ever before and allows for the exchange of more information. China has millions of Internet users, and although the government keeps a tight rein on free speech, the country has shown some signs

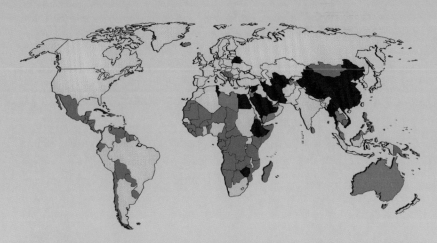

Internet Censorship Worldwide

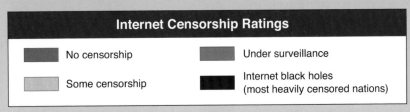

Internet Censorship Ratings

No censorship

Under surveillance

Some censorship

Internet black holes
(most heavily censored nations)

Source: Reporters Without Borders. Taken from: Internet Censorship: An Introductory Article, http://tech.dna24.co.cc/2009/05/27/internet-censorship-an-introductory-article/256/.html.

Internet cafes in some countries, like this one in Beijing, China, are monitored by the government, which blocks certain Web sites and monitors e-mail.

of allowing its citizens more Internet access. In the summer of 2009, the government decided to place censorship software only on computers in public places, not computers in private homes. Advancements in technology will also make it difficult to maintain a firm grip on all Internet speech. "Even with an army of censors, thought to number in the tens of thousands, monitoring computers in every home and 113,000 Internet cafés in China, it's difficult to control a medium as vast and mutable as the web,"[45] noted Clay Chandler in *Fortune* magazine.

Making an Impact

The Internet does offer hope to those determined to make their voice heard. In Burma (Myanmar) Internet café bloggers had a role in getting news out during an uprising against the government. Some news of the uprising came out through bloggers working in Internet cafés in Yangon who figured out how to get through the Internet-blocking software of the government before the government cracked down on the bloggers.

Sometimes the power of the Internet can allow a single person to make an impact by getting information out of a country with an oppressive government. Yoani Sanchez offers information about Cuba on her blog, Descuba.com. Writing on a blog hosted by a Web site that is outside of her country, she talks about everyday life in Cuba and offers criticism of the country's government. The Internet has given her a platform for making comments that traditional journalists in her country would not be able to make. "Under the nose of a regime that has never tolerated dissent, Sanchez has practiced what paper-bound journalists in her country cannot: freedom of speech,"[46] Oscar Hijuelos wrote in *Time*.

Some bloggers in Arab countries have also managed to use the Internet to make their opinions known. The magazine the *Economist* noted that young Arabs are blogging and using sites

Self-Censorship in Russia

When looking at the state of human rights in Russia, a U.S. report painted a gloomy picture in the area of Internet speech. "As some print and Internet media reflected a widening range of views, the government restricted media freedom through direct ownership of media outlets, pressuring the owners of major media outlets to abstain from critical coverage, and harassing and intimidating journalists into practicing self-censorship," the 2008 Human Rights Report: Russia from the U.S. State Department said.

Internet filtering could become another problem for Russians look-ing to share information online. The blog from the Internet and Democracy Project at Harvard University noted that in October 2009 the leader of Russia's Ministry of Communications supported filtering the Internet for negative content in order to protect children, and that a proposed law would require users to prove how old they are before being allowed to access certain sites.

U.S. Department of State, Bureau of Democracy, Human Rights and Labor, "2008 Human Rights Report: Russia," February 25, 2009. www.state.gov/g/drl/rls/hrrpt/2008/eur/119101.htm.

such as Facebook and YouTube to post information with political themes. Saudi Arabian women are using the Internet to protest their lack of rights in that country. When Saudi Arabia sent a team to the Beijing Olympics in 2008 that was composed only of men, images of the protest this action sparked were posted online.

Organizing Protestors

The Internet has also been used to help protestors organize and exercise free speech in other ways. In Egypt the April 6 Youth Movement evolved from a Facebook page. The page was founded to support striking workers in an industrial town, and Facebook and other social media tools were used to report on the strike. The site endured after the strike, growing to thousands of users, and was used to organize rallies and protests.

BYPASSING REPRESSION

"In repressive countries, the arrival of the internet has often been greeted by citizens as a wonderful way to bypass government control, enabling dissidents to air their views and keep in contact with the outside world."—The *Economist*.

Economist, "The Tongue Twisters," October 13, 2007, p. 66.

The site also illustrates the free-speech challenges people in Egypt still face, however. Several group members were arrested, and Facebook was used by the government as it worked to stop the protestors. The site helped lead them to the people organizing the protest.

Twitter and Iran

People continue to find new ways to use the Internet to advance their ideas. The ability to post information on the Internet site Twitter (which allows people to broadcast messages of 140 characters or less over the Web or through text messages) allowed

Some governments censor the official news, but users who post pictures and videos on sites like YouTube and Twitter allow citizens to receive the uncensored information.

protestors during the June 2009 Iranian presidential election to let their voices be heard.

The protestors were disputing the results, which had President Mahmoud Ahmadinejad winning almost 100 percent of the vote. As protests mounted the day after the election, the Iranian government tried to stop the protestors. Citizens and security forces clashed; hundreds were arrested and people were killed. Although the government kept tight limits on official news that was reported on the subject, protestors began sending messages on Twitter.

A video showing the shooting of a young woman named Neda was especially chilling and turned her into an icon against violence around the world. "The Twitterverse exploded with tweets in both English and Farsi," noted Lev Grossman in *Time* magazine. "While the front pages of Iranian newspapers were full of blank space where censors had whited out news stories, Twitter was delivering information from street level in real time."[47]

INTERNET CRACKDOWN

"Governments still fear dissenting opinion and try to shut it down. While the internet has brought freedom of information to millions, for some it has led to imprisonment by a government seeking to curtail that freedom. They have closed or censored websites and blogs; created firewalls to prevent access to information; and restricted and filtered search engines to keep information from their citizens."—Kate Allen, the United Kingdom director of Amnesty International.

Kate Allen, "Today, Our Chance to Fight a New Hi-tech Tyranny," *Observer*, May 28, 2006. www.guardian.co.uk/technology/2006/may/28/news.humanrights1.

The use of the Internet as a means of advancing free speech gained credibility when the U.S. State Department asked Twitter to delay a planned upgrade to its network until after the protests had cooled down, so the site would not be down even temporarily. The site was so important to the protestors that the United States did not want it shut down, lest the flow of information be interrupted. Iran's government controls much of the information presented on the country's radio and television stations and has the power to censor its newspapers, and Twitter offered an opportunity for the protestors to get their information out without censorship.

Limitations on Internet Speech

While Twitter was a wonderful method of getting information out during the elections in Iran, in a situation where voices would otherwise be silenced, the system was not perfect. The social network allowed people to get news out, but was criticized because so many comments were made that it became difficult to sift through them to find the best pieces of information. Although the Twitter posts allowed people to learn about what was going on, the comments did not offer the balanced and thorough reporting that a traditional news story would have presented.

Relying on social networking sites as a source of news can also be limited when governments slow Internet connections or block access to sites. Governments can also turn the tables on those promoting free speech online and use the Internet to send their own messages of propaganda.

Online Tools for Freedom of Speech

While some governments are trying to use technology to halt free speech on the Internet, others are doing the opposite. The same agency that runs Voice of America is testing technology that would let residents get through screens designed to limit their access to news. The news feed would be sent through e-mail accounts and is targeted at China, Iran, Vietnam, Myanmar (Burma), Uzbekistan, and Tajikistan.

In addition, the United States supports a service that allows users overseas to access almost any site on the Internet. "We

Amnesty International is a proponent of Internet freedom for all people, no matter what country. Pictured here, AI's Web site has been blocked from a computer in China.

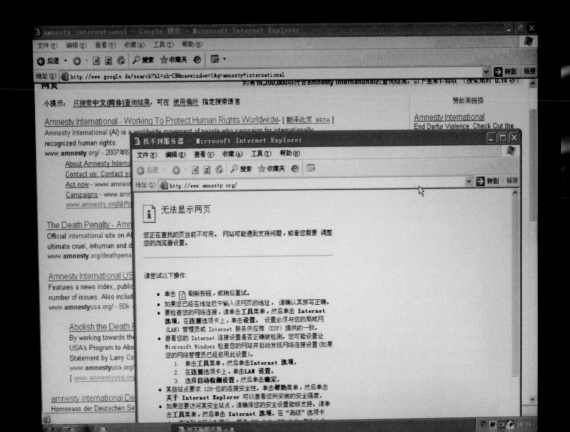

don't make any political statement about what people visit," said Ken Berman, head of information technology at the Broadcasting Board of Governors. "We are trying to impart the value: 'The more you know, the better.' People can look for themselves."[48]

NEW CHALLENGE

"The free flow of information has always threatened repressive regimes, and dictators have always sought to restrict it. The Internet is a new challenge for these regimes because it empowers with information and speech any person who can access the network."—Arizona senators John McCain and Jon Kyl.

John McCain and Jon Kyl, "The Internet and the 'Axis of (Censored),'" *USA Today*, February 21, 2006, p. 19A.

Software programmers around the world are also working to create free online tools that help people get around online censors who block Web sites or have access to e-mails. Tools for Internet free speech include encryption software, proxy servers, software that lets people remain anonymous, and social networking platforms that are secure. Hackers in countries where freedom of expression is restricted try to get around such restrictions by connecting to U.S. filters or using networks based in the United States to send anonymous messages.

One organization backing free speech around the world is Amnesty International. It is calling for people to take a stand for basic human rights by allowing for Internet freedom through its Irrepressible.info Campaign. Its Web site gives bloggers the opportunity to publish information that is censored in other countries. It also offers news on censorship issues around the world.

Libel Tourism

Some nations are using the Internet to try to impose their more restrictive laws on U.S. citizens. Americans are protected in their own borders against libel and slander, but because the Internet allows their works to travel around the world, they may find themselves being sued under another nation's laws. Under a trend called libel tourism, people who believe they have been defamed by an American author are taking their claims to court outside the United States.

In the United States people are protected by their First Amendment right to free speech. Libel claims are balanced with the belief that "public issues should be uninhibited, robust, and wide-open,"[49] as the Supreme Court stated in the landmark *Times v. Sullivan* case, which set the standard for determining whether or not a report about a public person can be considered defamatory. People claiming they have been defamed must prove that a person published a false statement of fact as a result of negligence or malice. Public figures claiming defamation must meet a higher standard than ordinary citizens and prove that the statement was published with actual malice, which means that the person knew the statement was false when it was published or published it with a reckless disregard for the truth.

Difficult Topic

The issue of libel tourism is such a touchy subject that one British journalist was advised to have his article on the subject carefully checked over by lawyers before it went to press. Oliver Marre, who writes for the *Observer* in London, said that British Member of Parliament Denis MacShane called him on his cell phone and warned, "Make sure your article is very carefully checked by lawyers before it is printed. Dealing with this subject is to swim in dangerous waters."

The sensitivity of the topic did not keep Marre from writing about it and sharing his opinion regarding the global impact of Britain's libel laws. "Free speech must not be put to flight by the approach of a bewigged British judge," he said. "Guarding and upholding the values of the U.S. Constitution is not just a worthy endeavor, but an essential safeguard to the American way of life, and something the rest of the world—most of all the UK government as they consider British libel laws—should learn from."

Quoted in Oliver Marre, "Suit Shopping: Is Libel Tourism a Threat to Free Speech—or Just to Neocons?" *American Conservative*, April 6, 2009. www.amconmag.com/article/2009/apr/06/00018.

Other countries do not view free-speech rights the same way America does. British law supports a person's reputation over the right to free speech. A person can be prosecuted for making a comment about a public official that is negligent or meant to defame that person. It is up to the person being sued to prove that the statement he made is true and not defamatory. The statement in question is presumed to be false unless the defendant can prove otherwise. People who live in Britain or do business there can claim to be libeled under British law. As a result, people who believe they have been defamed may bring up a case against an American in a British court.

Rachel Ehrenfeld, director of the American Center for Democracy, had firsthand experience with the implications of libel tourism. She was sued for libel by Saudi billionaire Khalid bin Mahfouz for linking him to terrorism in her book *Funding Evil: How Terrorism Is Financed and How to Stop It*. She claimed that he transferred $74 million to charities that gave funds to radical Muslim organizations such as al Qaeda and Hamas. Mahfouz sued Ehrenfeld for libel in England, although the book was not published or marketed there. However, some copies were sold over the Internet and shipped to people in England. Ehrenfeld chose not to challenge the suit Mahfouz brought against her, saying that she did not go to England because she did not recognize the country's legal authority over her. "I should not have to defend myself abroad,"[50] she said. A British judge ruled against her, giving her a fine and ordering her to destroy unsold copies of her book, but Ehrenfeld did not pay the fine or destroy her books.

Rachel's Law

Ehrenfeld campaigned for a change in American law that would prevent authors from being sued as she was. In 2008 the New York State Legislature passed Rachel's Law. The Libel Terrorism Protection Act allows New York courts to have legal authority over foreign libel litigants who have obtained a foreign defamation judgment against a New York writer or publication. Under the law, foreign libel judgments cannot be enforced by New York courts, unless the country where the judgment was decided meets or exceeds the U.S. standards for freedom of speech. It also gives a person the ability to have a court declare a foreign libel judgment to be invalid in New York. Similar acts were passed in Illinois and California.

Ehrenfeld's lawyer, Mark Stephens, supported a similar law on a national level. "[The law] means that libel tourists will not be able to use their wealth and power against people who cannot afford to defend themselves in Britain," he said. "It should be the case that people should sue in the jurisdiction where an alleged libel is published."[51]

WORLDWIDE IMPLICATIONS

"What we do in this country in terms of privacy and free speech isn't just about us. It's about a lot of people in the world who depend on our services or on the examples that our laws set."—Lee Tien, senior staff attorney for the Electronic Frontier Foundation.

Lee Tien, interview with author, August 11, 2009.

The Free Speech Protection Act, as proposed in 2009, would keep U.S. courts from enforcing judgments brought against Americans in foreign courts if what they wrote was not libelous under U.S. law. It would also allow Americans to countersue if their material is protected under the First Amendment and would give juries the option to award triple the amount of damages if the person bringing the foreign lawsuit was "intentionally engaged in a scheme to suppress rights under the First Amendment."[52]

Controversial Topic

Finding a way to protect the right of U.S. citizens to bring critical information about others to light, no matter how wealthy or powerful those being criticized are, is not an easy issue to solve. Preserving First Amendment rights is an important American value, but questions arise when it infringes on the rights of others to preserve their reputation or have due process of law. The issue is one that provides more fuel for the debate over free Internet speech. The Internet has made it possible to transfer information and goods easily around the world, but it has also brought new implications for questions over free-speech rights.

ADVANCING FREE
SPEECH

Although the information shared online is protected by the Constitution, Internet free-speech issues remain controversial. There are questions over who controls online content and whether further regulation is necessary. Concerns arise when Internet service providers take down information that has been posted. As the Internet evolves, court decisions, government regulations and Internet service providers have an impact on the state of online speech.

When the Internet was introduced, it was seen as an exciting way to post information without going through gatekeepers who controlled what was broadcast or published. Newspaper editors or video journalists controlled what was imparted over their news channels, but had no control over this new information medium. Mike Godwin compared the Internet to a town hall meeting "in which everyone has a chance to speak, no one is shouted down and everyone has time to explain his or her ideas. . . . The filtering function performed by newspaper editors is left to the readers, who are also contributors. . . . The very distinction between reader and reporter is blurred."[53]

New Gatekeepers

Because not everyone has his or her own Web site, however, other organizations have taken on some gatekeeping roles. When a video is posted on YouTube or a profile is created on Facebook, the person contributing the information does so under the Web site's rules. Companies such as YouTube, Facebook, MySpace, and other Web hosts have guidelines for users that outline what can be posted on their sites. Items can be taken down because of obscenity, copyright issues, excessive violence, nudity, or because they are threatening. Those who post material on these sites agree to abide by the sites' rules and regulations in return for the ability to post information.

Because of these rules people who visit the site can go there without stumbling on offensive material. Parents who let their children access the sites have some assurance that inappropriate content is being monitored. However, some are concerned that some Web site standards for removing content could silence a

Many Web hosts have guidelines that their users must accept and agree to in order to use their Web service.

person's ability to express political views. Exactly what will remain online and what will be removed is up to the company, which worries some advocates of free speech. "Those sites are where you'll find many of today's hottest items of political expression—beyond the reach of old rules and regulations," noted writer Ben Smith. "As political communication moves online, a new group of corporate executives and their customer service agents have gained control of the censor's pen—in a forum where the First Amendment does not apply."[54]

TAKEDOWN DECISIONS

"Attacks on individuals—groups saying 'Ben Smith is evil'—would get taken down. . . . But if it's 'Hillary Clinton is evil' or 'Rudy Giuliani is evil,' that's just fine with us, because we think that's core political speech."—Chris Kelly, Facebook's chief privacy officer.

Quoted in Ben Smith, "Internet Vulnerable to Free Speech Issues," Politico, May 9, 2007. www.politico.com/news/stories/0507/3919.html.

Smith pointed out that YouTube temporarily removed a video of a conservative blogger who made comments about Islam and that YouTube also took down a video of presidential candidate John McCain singing a few lines of a song that talked about bombing Iran. The videos were put back online, but Smith remained concerned about YouTube's ability to use vague guidelines to remove a video from view, even for a few hours.

Web site editors remove content for a number of reasons, including copyright violations and obscenity issues. What is taken down is ultimately up to the company, which must balance free speech with a desire to keep inappropriate material off the site. "We're all about allowing freedom of expression and allowing people to come into this new town hall," said YouTube editor Steve Grove. "But if somebody's going to say something that constitutes hate speech or doesn't meet our terms of service, we're going to take it down." He admitted that what is taken down is a matter of opinion, noting that "'inappropriate' is something of a vague term. We're a human company. Obviously humans have to make a decision at some point. It's imperfect by design, but we try to do our best."[55]

Net Neutrality

Another online issue that has been raised is whether Internet service providers should have control over how quickly information is sent. The companies that own the cables that transmit Internet information have the power to determine what gets sent first and which messages are sent fastest. They can also place limits on the size of e-mail messages. The concept of who has control over the speed of Internet content and whether certain applications can be blocked is known as Net neutrality. It becomes a freedom-of-speech issue when it relates to a person's right to have access to cyberspace.

Military Bloggers

Another place that free-speech rights come into question is the military. People in the military do not have the same freedom-of-speech rights while they are serving, since induction into the military indicates a change in status. Jason Hartley, an army reservist, was ordered to shut down his blog, Just Another Soldier, after posting comments that put down army life. When he first complied and then resumed posting, he was demoted to specialist because he had defied a direct order.

Free-speech rights in the military are curtailed because information provided in an online blog could be used by an enemy. In addition, dissension could undermine morale and weaken a unit. As in a school setting, limits on speech are necessary to maintain order and discipline. "We create cohesive teams of warriors who will bond so tightly that they are prepared to go into battle and give their lives if necessary for the accomplishment of the mission and for the cohesion of the group and for their individual buddies. We cannot allow anything to happen which would disrupt that feeling of cohesion within the force," said retired general Colin Powell.

Quoted in Frederick D. Thaden, "Blogs v. Freedom of Speech: A Commander's Primer Regarding First Amendment Rights as They Apply to the Blogopshere," *Reporter*, June 2006, p. 19.

To make sure that all Internet traffic is treated equally, the FCC has proposed rules that would make sure information flows freely over the Internet and that new applications are not blocked or slowed down. The rules would make sure that content such as a television show would not be slowed down if it was transmitted online. Those in favor of the regulations saw them as having the ability to prevent cable and telephone companies from blocking new Internet applications. However, those against the legislation say Net neutrality would make it more difficult for Internet service providers to control the traffic on their networks.

IS VENTING BAD FOR YOU?

"Net venting sites . . . can sometimes provide truthful, accurate and extremely helpful information that shames the wicked, deters the negligent, informs the naïve and warns potential victims. But the absence of editorial standards, the opportunity to post anonymously and the chance to 'get back' at someone for a petty oppression offer limitless opportunities for false, malicious and malignant postings, as well as a coarsening of society."—David A. Furlow, a Houston attorney whose practice involves First Amendment speech issues. He is a former Harris County assistant district attorney, and chaired a media, defamation, and privacy law committee for the American Bar Association.

David A. Furlow, "Net-Venting: Should a Server or a Speaker Face Civil Liability for Spite Speech on the World Wide Web?" *Andrews Litigation Reporter*, September 2007, p. 3. www.tklaw.com/resources/documents/PRV0501_FurlowComm.pdf.

The issue of Net neutrality has also been brought before Congress several times in the form of the Internet Freedom Preservation Act. The act would ensure that access to the Internet is equal. The bill was seen by some as a way to preserve an open Internet that gave everyone in America the opportunity to voice their opinion online and would allow for greater competition among those providing Internet access. However, concerns arose over the possibility of additional government control over the Internet. James Lakely, managing editor of *Infotech & Telecom News*, said that consumers already have the right to switch to a different Internet service provider if they do not like the one they are using, and this lets the industry regulate itself. "[Internet service providers] have an enormous financial incentive to retain existing customers and attract new ones, so the free market encourages best practices,"[56] he said.

Spam or Free Speech?

A person has the freedom to send information online, but spam has posed new issues, as it can clog in-boxes, contain viruses and bugs, and slow down personal computers. Unwanted e-mail can be difficult and costly to limit. George Pike, a writer for *Information Today*, notes that 90 percent of all e-mail traffic is spam and that this bulk e-mail can cost

Spam is unwanted e-mail that can clog in-boxes and infect computers with attached viruses.

businesses up to $17 billion in lost productivity and efforts to stop it. "The law's response to the proliferation of spam has been checkered," Pike said. "As the problems of spam shifted from its origins in the USENET environment in the 1980s and early 1990s to the email environment by the late 1990s, courts and legislatures struggled to craft a response. Unfortunately, that struggle continues today."[57]

Spam is controlled by the federal CAN-SPAM Act, passed in 2004. The act focuses on commercial spam and makes sure that commercial e-mail messages have accurate subject lines, headers, and sender addresses. While the act did not completely keep unwanted messages from appearing in people's inboxes, it did place some limits on the nuisance e-mails.

THE DEBATE CONTINUES

"The fight about free speech offline is still going on, I don't see why it would be different online."—Chris Hansen, senior staff attorney with the American Civil Liberties Union.

Chris Hansen, interview with author, August 10, 2009.

In 2009 spam regulation took another step when a bill to extend the CAN-SPAM regulations to mobile text messages was introduced. Mobile spam, also known as m-spam, involves sending unwanted text messages to a mobile phone. "Mobile spam invades both a consumer's cell phone and monthly bill," said Senator Olympia Snowe of Maine, who cosponsored the bill. "There is also increasing concern that mobile spam will become more than an annoyance—the viruses and malicious spyware that are often attached to traditional spam will most likely be more prevalent on wireless devices through m-spam."[58]

A Good Idea?

The content of some Internet speech is also cause for debate. While the right to post strongly critical opinions online has

This light-hearted computer keyboard reminds people to consider the consequences of their comments before posting them on the Internet.

been established, whether it is a good idea to make spiteful and nasty comments is debatable. The Internet is filled with lively commentary that can degrade into bitterness and ranting. Some Web sites allow people to criticize other drivers, talk about why they hate their job, and reveal how a boyfriend or girlfriend has cheated on them.

These comments may be legitimate free speech, but the wisdom of airing them in public is disputable. David A. Furlow, a Houston lawyer and former chair of the American Bar Association's media, defamation, and privacy law committee, said:

> Those who support postings on spite sites often say they enable ordinary people to "vent," releasing feelings of frustration, anger and inadequacy. However, it is questionable whether such tattling improves the blog poster's long-term mental health, avoids road-rage, cures broken hearts or keeps car tires from getting slashed. Instead of helping people move past unhappy events to get on with their lives, such postings can lead to retaliation, litigation and obsessions with past victimization.[59]

Criminal activity has been another source of debate on Internet speech. In Florida a 2008 law prohibited promoting gang activity on the Internet. Those who violated the law faced five years in prison. After the law was enacted, fourteen people associated with the Latin Kings gang were arrested in November 2008 in an effort that the Lee County sheriff's office called "Operation Firewall." One member allegedly used his MySpace page to post a hit list of people he wanted to kill, while another was shown online wearing gang colors and making a gang hand signal. In court the prosecutor argued that online expression intended to promote gangs was not protected because it was linked to criminal activity, but the defense attorney said the law the gang members were arrested under was unconstitutional. The court case remains unsettled but opens up a number of free-speech questions, noted Eric Robinson, a lawyer for the Media Law Resource Center. He said:

> While gangs and gang activity are serious problems, the Florida statute and the prosecutions under it are also problematic. Does displaying gang colors and symbols rise to the level of a credible threat of violence or solicitation to imminent criminal activity? Is an online "hit list" meant to be taken seriously? And what are the implications of this type of prosecution for free speech, both online and off?[60]

Continuing Debate

The debate over free speech is far from over. Questions persist over the need for regulations or limits and the impact these would have on Americans' First Amendment rights. Some see government oversight as a way to make sure all Internet users are treated equally, while others see this as an intrusion on a free market. New communication methods, from mobile phones to Twitter applications, also bring new challenges as they take the Internet out of people's homes and make the Internet a mobile communications medium.

The controversies do not surprise Lee Tien, senior staff attorney at Electronic Frontier Foundation, which has fought since the 1990s for Internet speech to be open and free. He said:

Wireless Control

While issues over whether wireless Internet control should be regulated by the government continue to be debated in the United States, the issue is not as controversial in other countries where a person can buy a wireless device and use it with any carrier. That allows any applications to be placed on the device. This would help consumers get the device, applications, and wireless service they want, said Eli Noam, director of Columbia University's Tele-Information Institute. "Carriers should not have gatekeeper control over applications," he said.

Quoted in Leslie Cauley, "Wireless Lockdown Catches FCC's Eye," *USA Today,* August 14, 2009, p. 1B.

Every media fights for its First Amendment identity. Every media gets placed into a public dispute about things like sex, things like defamation, things like cyberstalking. There are all sorts of things that will always be used to say this place is too freewheeling, we need to place some disciplinary controls on it. Our role is to fight censorship, how dangerous censorship is to freedom of speech and how easy it is for speech to be chilled.[61]

Introduction: A New Era for Free Speech

1. Quoted in FindLaw for Legal Professionals, "New York Times Co. vs. Sullivan." http://caselaw.lp.findlaw.com/scripts/getcase .pl?court=US&vol=376&invol=254.

Chapter 1: Free Speech and Responsibility

2. Quoted in "Testimony of Ben Scott Policy Director Free Press before the U.S. House of Representatives Subcommittee on Telecommunications and the Internet of the Committee on Energy and Commerce on behalf of the Free Press Consumers Union Consumer Federation of America Public Knowledge regarding A Legislative Hearing on H.R. 5353, The Internet Freedom Preservation Act of 2008," freepress, May 6, 2008, www.freepress.net/files/Scott_Testimony_FINAL_ 5-6-08.pdf.

3. Mary-Rose Papandrea, interview with author, September 16, 2009.

4. Quoted in Citizen Media Law Project, "Bell v. Shah," September 10, 2007. www.citmedialaw.org/threats/bell-v-shah.

Chapter 2: Open Online Dialogue

5. Lee Tien, interview with author, August 11, 2009.

6. Chris Hansen, telephone interview with author, August 10, 2009.

7. Hansen, interview.

8. Justice John Paul Stevens, "Janet Reno, Attorney General of the United States, et al., Appellants v. American Civil Liberties Union et al.," Cornell University Law School Supreme Court Collection. www.law.cornell.edu/supct/html/96-511 .ZO.html.

9. John W. Kennedy, "Child Online Protection Act Challenged," *Christianity Today*, December 7, 1998, p. 19.

10. *New York Times*, "A Win for Free Speech Online," January 27, 2009, p. 30.

11. George H. Pike, "A Safe Harbor Against Lawsuits," *Information Today*, November 2006, p. 17.

12. Tien, interview.

13. Tien, interview.

14. Quoted in Citizen Media Law Project, "Cubby v. Compuserve," October 15, 2007. www.citmedialaw.org/threats/cubby-v-compuserve.

Chapter 3: Who Said That?

15. George H. Pike, "The Right to Remain Anonymous," *Information Today*, April 2008, p. 15.

16. Barry Bruce, "The Big Chill," *Publisher's Weekly*, May 21, 2007, p. 62.

17. Pike, "The Right to Remain Anonymous," p. 15.

18. Quoted in Citizen Media Law Project, "Cahill v. Doe (Schaeffer)," September 20, 2007. www.citmedialaw.org/threats/cahill-v-doe-schaeffer.

19. Quoted in Asher Moses, "Model Forces Google to Reveal 'Skank' Blogger's Identity," *Sydney Morning Herald*, August 19, 2009. www.smh.com.au/technology/technology-news/model-forces-google-to-reveal-skank-bloggers-identity-20090819-epz0.html.

20. Quoted in George Rush, "Outed Blogger Rosemary Port Blames Model Liskula Cohen for 'Skank' Stink," *New York Daily News*, August 23, 2009. www.nydailynews.com/gossip/2009/08/23/2009-08-23_outted_blogger_rosemary_port_blames_model_liskula_cohen_for_skank_stink.html.

21. Quoted in Sean Woods, "Big Brother vs. Blogs," *Rolling Stone*, May 3, 2007, p. 40.

Chapter 4: Copyright and Criticism

22. Chuck Leddy, "Copyright Piracy Grows with New Technology," *Writer*, September 2007, p. 8.

23. Leddy, "Copyright Piracy Grows with New Technology," p. 8.

24. Quoted in Grant Gross, "Official: Book Settlement Makes 'Mockery' of Copyright Law," SFGate, September 10, 2009. www.sfgate.com/cgi-bin/article.cgi?f=/g/a/2009/09/10/urn idgns002570F3005978D88525762D0054575D.DTL.

25. Quoted in Juan Carlos Perez, "DOJ: Court Should Reject Google Book Search Settlement," SFGate, September 18, 2009. www.sfgate.com/cgi-bin/article.cgi?f=/g/a/2009/09/18/urnidgns002570F3005978D8852576340079E02E.DTL.

26. Quoted in Electronic Frontier Foundation, "Online Policy Group v. Diebold," www.eff.org/cases/online-policy-group-v-diebold.

27. Tim Jones, "YouTube Restores a Fair Use," Electronic Frontier Foundation, May 7, 2009. www.eff.org/deeplinks/2009/05/youtube-restores.

28. Quoted in Electronic Frontier Foundation, "Parody Website Back Online After Settlement of Bogus IP Claims," February 2, 2009. www.eff.org/press/archives/2009/02/02.

29. Quoted in Electronic Frontier Foundation, "Parody Website Back Online After Settlement of Bogus IP Claims."

Chapter 5: School Speech and Cyberbullies

30. Carol Brydolf, "Minding MySpace: Balancing the Benefits and Risks of Students' Online Social Networks," *Education Digest*, October 2007, p. 4.

31. Papandrea, interview.

32. Quoted in Mark Hamblett, "2nd Circuit Upholds Student's Suspension for Instant-Messaging Violent Image," Law.com, July 6, 2007. www.law.com/jsp/article.jsp?id=1183626396604.

33. Quoted in Brian Stewart, "Student Files Lawsuit After Coach Distributed Private Facebook Content," Student Press Law Center, July 22, 2009. www.splc.org/newsflash.asp?id=1938.

34. Quoted in Mark Walsh, "Student Loses Discipline Case for Blog Remarks," *Education Week*, June 11, 2008, p. 7.

35. Quoted in Walsh, "Student Loses Discipline Case for Blog Remarks," p. 7.

36. Papandrea, interview.

37. Edwin C. Dardin, "The Cyber Jungle," *American School Board Journal*, April 2009, p. 55.

38. Brydolf, "Minding MySpace," p. 4.

39. Quoted in Reid J. Epstein, "Oceanside Teen Sues Classmates, Facebook for $3M," *Newsday*, March 2, 2009. www.newsday.com/long-island/nassau/oceanside-teen-sues-facebook-ex-classmates-for-3m-1.896009.

40. David Kravets, "Student Who Created Facebook Group Critical of Teacher Sues High School Over Suspension," *Wired*, December 9, 2008. www.wired.com/threatlevel/2008/12/us-student-inte.

41. Quoted in Lauren Collins, "Friend Game," *New Yorker*, January 21, 2008, p. 34.

42. Quoted in Maggie Shiels, "Cyber Bullying Case Sentence Due," BBC News. http://news.bbc.co.uk/2/hi/technology/8127533.stm.

43. Quoted in Shiels, "Cyber Bullying Case Sentence Due."

44. Quoted in Shiels, "Cyber Bullying Case Sentence Due."

Chapter 6: International Free Speech

45. Clay Chandler et al., "Inside the Great Firewall of China," *Fortune*, March 6, 2006, p. 148.

46. Oscar Hijuelos, "Yoani Sanchez," *Time*, May 12, 2008, p. 68.

47. Lev Grossman, "The Moment," *Time*, June 29, 2009. www.time.com/time/magazine/article/0,9171,1905527,00.html.

48. Quoted in Jim Finkle, "U.S. Tests Technology to Break Through Foreign Web Censorship," Reuters. www.reuters.com/article/newsOne/idUSTRE57D14R20090814.

49. Quoted in "U.S. Supreme Court New York Times Co. v. Sullivan, 376 U.W. 254 (1964)," Justia.com, http://supreme.justia.com/us/376/254/case.html.

50. Quoted in Rachel Ehrenfeld, "Libel Tourism," FDCH Congressional Testimony, February 12, 2009. http://search.eb

scohost.com/login.aspx?direct=true&db=n5h&AN=32Y21
61944422&loginpage=Login.asp&site=ehost-live.

51. Quoted in Roy Greenslade, "An End to the Libel Tourist
Trap," *Guardian*, October 20, 2008. www.guardian.co.uk/
media/2008/oct/20/pressandpublishing1.

52. Quoted in GovTrack.us, "S. 449: Free Speech Protection Act
of 2009." www.govtrack.us/congress/bill.xpd?bill=s111-449
&tab=summary.

Chapter 7: Advancing Free Speech

53. Mike Godwin, *Cyber Rights: Defending Free Speech in the Digital Age*. New York: Times Books, 1998, p. 9.

54. Ben Smith, "Internet Vulnerable to Free Speech Issues," Politico, May 9, 2007. www.politico.com/news/stories/0507/
3919.html.

55. Quoted in Smith, "Internet Vulnerable to Free Speech Issues."

56. James G. Lakely, "'Net Neutrality': There Goes the Neighborhood," *Bulletin*, August 4, 2009. http://thebulletin.us/
articles/2009/08/04/commentary/op-eds/doc4a7875
be95a23343495891.txt.

57. George Pike, "Anti-spam Legislation Setbacks," *Information Today*, December 2008, p. 17.

58. Quoted in Kristina Knight, "Can-SPAM, Now Can-m-SPAM,"
BizReport, April 7, 2009. www.bizreport.com/2009/04/
can-spam_now_can-m-spam.html.

59. David A. Furlow, "Net-Venting: Should a Server or a Speaker Face Civil Liability for Spite Speech on the World Wide Web?" *Andrews Litigation Reporter*, September 2007, p. 3.
www.tklaw.com/resources/documents/PRV0501_Furlow
Comm.pdf.

60. Eric P. Robinson, "Florida Sees Gangs in Social Networks,
Prosecutes," Citizen Media Law Project, August 6, 2009.
www.citmedialaw.org/blog/2009/florida-sees-gangs-social-
networks-and-prosecutes.

61. Tien, interview.

Chapter 1: Free Speech and Responsibility

1. Should a student be punished for creating a fake profile of a teacher on a social networking site? Why or why not?

2. In your opinion, when does criticism of someone become defamation?

3. Why is it dangerous to post untrue statements on the Internet?

Chapter 2: Open Online Dialogue

1. Do you think Internet service providers should have the protection they were given under Section 230? Why or why not?

2. What restrictions would you place on online speech?

3. Should obscene words be allowed online? Why or why not?

Chapter 3: Who Said That?

1. What are the pros and cons of allowing people to remain anonymous online?

2. Why would a person file a John Doe lawsuit?

3. When should an Internet service provider reveal an anonymous person's name?

Chapter 4: Copyright and Criticism

1. How would you define "fair use?"

2. Do you think the employee who used photos of Nazis to criticize his employer was treated fairly? Why or why not?

3. How can copyright law be wrongly used to stop a person from offering criticism?

Chapter 5: School Speech and Cyberbullies

1. Should students be punished for what they say online when they are not in school? Why or why not?

2. Do you feel a student's right to free speech is unfairly restricted? Why or why not?

3. What can be done to prevent cyberbullying?

Chapter 6: International Free Speech

1. Should people in countries that have laws restricting free speech break the law to have their voice heard? Why or why not?

2. Why would a government restrict free speech?

3. How could you encourage free speech in a foreign country?

Chapter 7: Advancing Free Speech

1. What do you see as the most important issue facing Internet free speech?

2. If you ran YouTube, what rules would you have for what can be posted and not posted? What would be your takedown policy?

3. Do you think Internet speech will be the same, more free, or more restricted in the future? Why?

ORGANIZATIONS TO CONTACT

American Civil Liberties Union
125 Broad St., 18th floor
New York, NY 10004
phone: (212) 549-2500
Web site: www.aclu.org

With more than five hundred thousand members, this organization works to defend the rights and liberties of Americans, including the First Amendment right to free speech. Its staff attorneys and volunteer attorneys handle many civil liberties cases each year.

Amnesty International
1 Easton St.
London
WC1X 0DW, UK
phone: 44-20-74135500
fax: 44-20-79561157
Web site: www.amnesty.org

This worldwide organization campaigns for human rights for all people. It raises awareness of human rights issues in countries around the world. It publishes the e-newsletters *Stop Violence Against Women* and *Counter Terror with Justice*.

Center for Democracy and Technology
1634 I St. NW, #1100
Washington, DC 20006
phone: (202) 637-9800
fax: (202) 637-0968
Web site: www.cdt.org

This nonprofit, public interest organization supports keeping the Internet open and free. Its Web site includes a list of reports and articles relating to that topic.

Citizen Media Law Project
Berkman Center for Internet and Society
Harvard Law School
23 Everett St., 2nd Floor
Cambridge, MA 02138
phone: (617) 495-7547
fax: (617) 495-7641
Web site: www.citmedialaw.org

This project hosted by the Berkman Center for Internet and Society at Harvard Law School provides legal assistance and resources for people involved in online media. It is also affiliated with the Center for Citizen Media.

Electronic Frontier Foundation
454 Shotwell St.
San Francisco, CA 94110-1914
phone: (415) 436-9333
fax: (415) 436-9993
Web site: www.eff.org

Founded in 1990, this organization addresses issues relating to online and digital rights. The Deeplinks blog on its Web site offers commentary on news items relating to free-speech issues.

Freedom House
1301 Connecticut Ave. NW, Floor 6
Washington, DC 20036
phone: (202) 296-5101
fax: (202) 293-2840
Web site: www.freedomhouse.org

This nonprofit organization supports democracy and freedom around the world. It was founded in 1941 and publishes several annual reports looking at freedom on a global scale. It publishes several annual surveys, including *Freedom of the Press*, which looks at media independence around the world.

International Federation of Journalists
IPC-Residence Palace, Bloc C
Rue de la Loi 155

B-1040 Brussels
Belgium
phone: 32-2-235 22 00
fax: 32-2-235 22 19
Web site: www.ifj.org

Dedicated to freedom of the press around the world, this organization monitors violations of press freedom and campaigns for the safety of journalists. It publishes reports, which are available on its Web site, that track the number of journalists and staff media that are killed each year. Its reports also look at the state of global press freedom.

OpenNet Initiative

23 Everett St., 2nd Floor
Cambridge, MA 02138
phone: (617) 495-7547
fax: (617) 495-7641
Web site: www.opennet.net

This organization is a collaborative partnership between the University of Toronto, Harvard University, the University of Cambridge, and Oxford University. Its mission is to identify and document Internet filtering and surveillance. It publishes online country profiles that look at a country's Internet filtering practices. Its Web site also contains reports and articles about recent events impacting the filtering of online information.

Society of Professional Journalists

Eugene S. Pulliam National Journalism Center
3909 N. Meridian St.
Indianapolis, IN 46208
phone: (317) 927-8000
fax:(317) 920-4789
Web site: www.spj.org

Founded in 1909, the organization upholds high ethical standards and is dedicated to protecting free speech and press freedom. The society offers training for citizen journalists, and its Web site contains articles relating to free speech and press freedom around the world.

Books

Laura Egendorf, ed., *Should There Be Limits to Free Speech?* San Diego: Greenhaven, 2003. A collection of commentaries and various viewpoints on free speech.

Laura Egendorf, ed., *Students' Rights.* Detroit: Greenhaven, 2006. Information and commentary on court cases that impact student rights.

Sherri Mabry Gordon, *Downloading Copyrighted Stuff from the Internet: Stealing or Fair Use?* Berkeley Heights, NJ: Enslow, 2006. This young adult book explains what is allowed and what is not under copyright law.

Periodicals

Joanne Leedom-Ackerman, "The Intensifying Battle Over Internet Freedom," *Christian Science Monitor*, February 24, 2009. www.csmonitor.com/2009/0224/p09s01-coop.html. Column on online free-speech concerns around the world.

Oliver Marre, "Suit Shopping: Is Libel Tourism a Threat to Free Speech—or Just to Neocons?" *American Conservative*, April 6, 2009. www.amconmag.com/article/2009/apr/06/00018. A British journalist discusses libel tourism.

Web Sites

Amnesty International (www.amnesty.org). Amnesty International protects human rights around the world and promotes Internet freedom through its Irrepressible.info Campaign.

Citizen Media Law Project (www.citmedialaw.org). Contains summaries and updated information on lawsuits relating to free speech.

Electronic Frontier Foundation (www.eff.org). Provides information on recent news relating to Internet free speech as well as blogger's rights, relevant court cases, and international events.

GovTrack.us (www.govtrack.us). Stays on top of where congressional bills are and can be used to find pending legislation impacting free speech.

Internet and Democracy Blog (http://blogs.law.harvard.edu/idblog/category/free-speech). The team blog for the Internet and Democracy Project at the Berkman Center for Internet and Society at Harvard University offers updates on Internet free-speech issues around the world.

Public Citizen (www.citizen.org/litigation/briefs/IntFreeSpch). The Web site of this public interest organization offers a guide for bloggers and information on Internet free-speech cases.

INDEX

PICTURE CREDITS

Cover: Image copyright Nagy-Bagoly Arpad, 2009. Used under license from Shutter stock.com

© AKP Photos/Alamy, 52

© Coffeehouse Productions/Alamy, 18

© Duffie/Alamy, 91

© David J. Green—lifestyle themes/Alamy, 37

© Juergen Hasenkopf/Alamy, 72

© ICP/Alamy, 85

© INSADCO Photography/Alamy, 26

© Oleksiy Maksymenko/Alamy, 8

© Nuage/Alamy, 36

© pixel.pix/Alamy, 28

© Alex Segre/Alamy, 16

© Jack Sullivan/Alamy, 25

© Rawdon Wyatt/Alamy, 62

© 1bestofphoto/Alamy, 79

AP Images, 66

© Bettmann/Corbis, 57

© Danny Lehman/Corbis, 74

© JULIAN STRATENSCHULTE/epa/Corbis, 77

© Larry Downing/Sygma/Corbis, 32

Image copyright Lim Yong Hian, 2010. Used under license from Shutterstock.com, 89

Image copyright David Hughes, 2010. Used under license from Shutterstock.com, 58

Image copyright Irucik, 2010. Used under license from Shutterstock.com, 41

Image copyright monared, 2010. Used under license from Shutterstock.com, 45

Image copyright Oldproof, 2010. Used under license from Shutterstock.com, 48

Image copyright ostill, 2010. Used under license from Shutterstock.com, 12

Image copyright Matthias Pahl, 2010. Used under license from Shutterstock.com, 11

Image copyright Losevsky Pavel, 2010. Used under license from Shutterstock.com, 23

ABOUT THE AUTHOR

Terri Dougherty has degrees in journalism and English and has written more than eighty books for children. She finds the topic of free speech fascinating. In addition to writing, she enjoys traveling and spending time with her husband, Denis, and their three children, Kyle, Rachel, and Emily.